Wild Africa's Silent Call

by the same author

TRAVEL AND FICTION

Tropical Quest (3rd impression)
Eastern Quest
The Motorist's Holiday Guide to Europe
The Photographer's Holiday Guide to Europe
Continental Autocamping
A Guide to Germany
Tag Along Tiger
A Night in Trinidad (B.B.C. Radio)
A Question of Psychology (B.B.C. Radio)
Townsend's Guide to the Caribbean
Let's Visit East Africa

PHOTOGRAPHIC

Underwater Photography
The Practical Guide to Holiday and Family Movies
Your A-Z Guide to Better Movies (also in paperback)
How to Use 16 mm. (also in paperback)
Filming in Colour (also in paperback)
Film in Research
Cine Photography (2nd impression)
Photographic Processing
Everyone's Guide to Movie Making
Sporting Photography
A Complete Guide to Home Movies
Glossary of Terms in Photography and Cinematography (four languages)
Photography in Colour (B.B.C. Radio)
Better Homes and Gardens – Photography for your Family (Editor)
Holiday Photography in Colour (in preparation)

Frontispiece: Jill meets Ziggy, a young ostrich, in the Kidepo Valley, Uganda

Wild Africa's Silent Call

A Quest through East Africa

by
DERAK TOWNSEND

with a Foreword by
PETER SCOTT

London
GEORGE ALLEN AND UNWIN LTD

FIRST PUBLISHED IN 1969

SBN 04 916007 9

PRINTED IN GREAT BRITAIN
in 11 on 12pt Baskerville type
BY C. TINLING & CO. LTD
PRESCOT

The wild life of today is not ours to dispose of as we please. We have it in trust, and must account for it to those who come after.

King George VI

Thank you

DURING such a long safari we are naturally indebted to many people throughout East Africa for their kind help and friendship. It would take many pages to acknowledge everyone concerned and rather than omit any name – either inadvertently or simply through lack of space – I prefer to say one big 'thank you'.

Likewise, numerous African business organizations offered us invaluable facilities and advice but although these are not mentioned in the text, it does not mean their great assistance has been overlooked. Indeed, I am truly grateful.

And for my readers, I would like to make it clear that none of our journeys are sponsored trips, for where direct sponsorship is concerned, an author is under an obligation which must inevitably influence his writing about the areas concerned. (This particular aspect of my true reporting has been covered more fully in pages 91 to 93).

However, within the story I do include a certain amount of deliberate 'product identification' because feature writers – especially for magazine articles – have always asked most specifically *what* we take with us, the kind of camera(s) I use, the brand of skin cream that Jill favours under tropical conditions, the make of typewriter I hammer, the type of watch I wear and a thousand other such queries. Many of our readers share the same interest and letters sent to us c/o Allen & Unwin often cover exactly the same questions, particularly from those who are planning to visit the country.

In this book, I have therefore given more practical information than usual regarding specialized equipment and where praise is due, praise has been given. Our hired Ford Cortina is featured most prominently but the Ford Motor Company had no idea that I was even in East Africa let alone driving one of their cars. As for our extensive use of the excellent Land

Rover – well, I doubt if the Rover Motor Company knows of my existence!

There is only one exception. A special acknowledgement I wish to give A. Wetter & Company of Zurich, Switzerland. This company is among the finest blockmakers in Europe and to help the World Wildlife Fund they kindly donated the expensive colour printing blocks for this publication.

CONTENTS

ILLUSTRATIONS

Colour Plates

Black and White

All photographs by the author.
Colour letterpress printing plates – A. Wetter & Company, 17 Hotzestrasse, Zurich 6, Switzerland.

Photographic processing – Quickprint Photographic Services Limited, Chartham Road, South Norwood, London, S.E.25.

For East Africa – for its ever changing spectacular landscape – for its warm hearted people – for its azure sky – for its fascinating wild life.

For those conservationists who fight a never ending battle to protect these wonders, to arrest the destruction caused by expanding settlement, to stop the ravages of animal hunters, to save the pastoral tribes from being relentlessly squeezed into extinction and to preserve as much as possible of the world in its entirety.

To all these great pioneers who defend nature at every opportunity, I dedicate *Wild Africa's Silent Call* most humbly.

The author has donated a percentage of his proceeds from the sale of this book to The British National Appeal of the World Wildlife Fund, 7–8 Plumtree Court, London, E.C.4.

Foreword

BY PETER SCOTT

'ARMCHAIR travellers' are fortunate that authors like Derek Townsend make it their business, with pen and camera, to recount their journeys throughout the world. Even with the massive communications systems of today, it is often through the written word that the most personal contact is made.

Wild Africa's Silent Call is an essentially personal account of Derek Townsend's safari through East Africa. His unobtrusively detailed narrative is illuminated by an intense interest in everything he sees and everyone he meets.

With the help of books such as this, people are beginning to value the disappearing natural world; to realize that the earth is small in comparison with the destructive power of modern man and the vast scale on which he can change (and much too often spoil) the natural scene.

Wild life and wild places have suddenly begun to seem valuable after all. Suddenly we have woken up to the carelessness which brought extinction to the Dodo, the Quagga, the Blue Antelope and the Passenger Pigeon. We recognize the tragedy of the deliberately burnt forests, the dust bowls, the erosion, the near extinction of the great whales – and all the other profligate things we have been doing to our earth in our ignorance.

We begin to think of what we should be doing for the conservation of nature which, of course, is mainly important because of what nature does for us.

I believe that something goes wrong with people when they cut themselves off from the natural world. This is why we keep gardens or window-boxes or house plants, or dogs or cats or budgerigars. Man does not live by bread alone, and never was this more true than today. I believe we should take just as great pains to look after the natural treasures which inspire us as we do to preserve the man-made treasures in art galleries and museums. This is a responsibility we have to future generations, just as we're responsible for the safeguarding of great works of art.

It has been argued that if the human population of the world continues to increase at its present rate, there will soon be no room for either wild life or wild places, so why waste time, effort and money trying to conserve them now? But I believe that sooner or later we shall learn to limit our own over-population which, in my view, is one of humanity's greatest dangers. We shall become concerned with optimum rather than maximum, with quality rather than quantity, and everyone will rediscover the need within themselves for contact with wilderness and wild nature as an essential part of a 'high quality environment' – the importance of surroundings in our lives, and the special significance of wildness.

Of course, no one can tell quite when that will happen, and I am concerned that when it does, breeding stocks of wild animals and plants should still exist, preserved perhaps mainly in nature reserves and national parks (even in zoos and botanical gardens) to repopulate the natural environment which I believe we shall then wish to recreate and rehabilitate.

These are my reasons for believing passionately in the conservation of nature.

All this calls for action of three main kinds: first more research in ecology – the science of the relationship between soil and water, and plants and animals and man himself – the science, if you like, of our natural environment. Second, the setting aside of more land as effective inviolate strongholds, nature reserves and national parks. Finally, and above all, education. By calling attention to the plight of the world's wild life, and by encouraging people to enrich their lives by the enjoyment of nature, it may be possible to accelerate the change in outlook and also to trigger off the necessary action.

It has been estimated that the conservation picture all over the world could be transformed by an injection of about £2 million each year, which is not an astronomical figure. Of course, a lot of money is needed for relieving human suffering, and charity for this work is obviously a high priority for all of us – for all people of goodwill and compassion; but some money is also needed for human fulfilment and inspiration.

Conservation, like education and art, claims some proportion of the money we give to help other people including, incidentally, those not yet born.

The World Wildlife Fund, with its international headquarters in Switzerland, is the community chest for all the conservation efforts in the world. It exists to provide the money which specialists need to save wild life and wilderness for the enjoyment of mankind now and in the future. It has already supported hundreds of projects each of which has been very carefully screened to make sure that it is scientifically sound and that none of the money is wasted. It needs, each year to do its job properly, about the price of a mile of motorway in flat country.

His Majesty King George VI once said: 'The wild life of today is not ours to dispose of as we please. We have it in trust, and must account for it to those who come after.'

In his travels, Derek Townsend has become more and more convinced of this truth and I hope that everyone who reads *Wild Africa's Silent Call* will recognize it, too.

Peter Scott

C.B.E., D.S.C., LL.D.

The Call of the Wilds

TODAY, there are few remaining accessible places where man can escape from the imprisonment created by his own technological progress. Stone, metal and cement fabrications devastate the world's surface. Industry pollutes the air we breathe. Skyscraper buildings keep the sunlight off our streets and blot out the stars. Indeed, these barriers so effectively isolate us from nature that many city children have never experienced even simple pleasures like smelling the early morning fragrance of flowers or the invigorating odour of damp earth.

Yet, although the majority of people have had little opportunity to develop their perception and appreciation, they still retain an affinity with the essential qualities of life, for the womb of Mother Nature is where all humanity was nurtured.

More than ever before, people need a retreat from their artificial environment, for the stimulus of the wilds awakens a new power of creation – it mellows – it soothes – it relaxes jangled nerves – it recharges the mental batteries.

And so I was excited, for having chosen to make our third quest through East Africa, the great outdoors promised not only panoramic skies, equatorial sunsets, remote forest camps, snow-capped mountains and vast tracts of countryside untrampled by civilization, but also it offered the thrill of living close to the greatest concentration of wild life to be found anywhere.

East Africa is a land where daylight brings the incessant barking of baboons, the vibrating rumble of elephant talk, the penetrating snort from a disturbed rhino, the grunt of an enraged buffalo and the whinny of zebras as they gather around a waterhole. It is a land where dusk falls to the piercing screech of a rock hyrax, the flutter of an eagle's wings or the full-throated roar of the lioness after making her kill. It is a land where darkness is disturbed by the hyena's maniacal laugh, the leopard's asthmatical wheeze and the 'haw-haw-

haw' bellow of hippopotami. It is a unique land, the Noah's ark of the twentieth century where the call of the wilds is urgent and compelling.

Perhaps those who can no longer recognize this call, whose perception has been dulled, may once again find their senses responding to the magnificence of untamed Africa which I shall do my best to recreate.

CHAPTER I

Tea with a Chimpanzee

No ships – Route hazards – Land Rovers and a Ford Cortina – Jill arrives in Nairobi – A vast continent – On our way to Arusha – Holes – Wildebeest make a thrilling introduction to wild Africa – Spear carrying warriors – The border post – Into Tanzania – Only a political wrangle – An unusual game sanctuary – Beasts and birds – Hell breaks loose – A lonely dinner – Giant snakes – A strange country.

SHIPS to East Africa, even to Mombasa and Dar es Salaam, are few and far between. In fact we found only five shipping lines sailing to either of these ports from the United Kingdom and, because the Suez Canal was still closed after the Israel-Arab conflict, the voyage was taking an average of thirty days around the Cape. Furthermore there were no vessels scheduled to leave for several weeks, which meant that an important part of our journey through Tanzania would be in the southeast monsoon period of long rains, making the study of animals very difficult and photography all but impossible.

We therefore scrapped our original idea of taking a vehicle with us by sea and decided to fly out and hire transport. But it was not just a matter of walking into any car rental firm, signing on the dotted line and driving off. Our intended route was far too extensive for such casual treatment.

Covering three countries over a period of many months, there would often be times when we were 200 miles from a petrol pump, and when a serious breakdown could mean waiting for days before anyone appeared on the scene to offer help. The roads themselves, twisting upwards from the hot, dry lowlands to the coolness of volcanic peaks at 10,000 feet or more, were going to provide extra obstacles; soft sand, mud, fissures, corrugated surfaces, potholes, floods and rocky tracks would all offer their individual hazards.

It was therefore essential to choose our vehicle with great care and to plan the expedition in detail, gathering vital information about such things as the climatic conditions likely to be experienced, where the wild life lodges were sited, the location of petrol stations, what provisions to carry and the driving procedure to adopt when encountering potentially dangerous animals like elephants. In order to make these arrangements and to save expense while doing so, I flew out to Nairobi a fortnight ahead of my wife Jill.

Before leaving London, the Kenya High Commission had kindly given me an introduction to UTC (United Touring Company) the biggest and most experienced safari operators in Africa. It was an invaluable recommendation and much of my fortnight was spent at the UTC offices where nothing seemed too much trouble for their staff. This organization was founded over twenty years ago with three cars; now they have a fleet of over 1,200 vehicles and even offer an air charter service.

At first I wanted to hire a four-wheel-drive Land Rover, but taking the advice of experienced safari men I settled for the rugged Ford Cortina, which together with the Peugeot 404 has proved itself so admirably in the East African Safari Rally. For certain locations only Land Rovers are practicable, but over such a long period UTC thought it would be less tiring and less expensive on fuel to treat these areas as separate expeditions and to hire Land Rovers locally. This proved to be very wise advice.

And so, in a white Cortina, I was awaiting Jill's arrival at Nairobi airport when her British United VC 10 made its approach run, the four rear mounted Rolls Royce Conway engines leaving behind light trails of smoke which scratched a grey course across the sky. The plane made a smooth landing, the engines blasted into reverse thrust only for seconds and then, exactly on time, the airliner came to a final halt outside the terminal building. I left the observation terrace and went to the arrivals lounge. In just a few more hours we would be setting off across the great wilderness of Masailand.

Early the next morning we drove out of Nairobi. At first the road was good, stretching out through the scrubland like a wide grey tape, although nearing the horizon it narrowed to a frail thread of silver and disappeared in a shimmering rivulet of heat.

Tall, slender, yellow-barked acacia trees with vivid green foliage and strange flat tops grew in scattered profusion, looking as if they had been scalped and pruned by a zealous gardener. Jill called them 'the flat-topped trees'.

Among the acacias stood huge mounds of red earth often about twelve feet high. Some of these mounds had delicately-fashioned spires and pinnacles like one of Ludwig's fairy tale palaces in Germany. Some exhibited rounded domes; others were shaped like an Egyptian pyramid. But each one had been created by the same skilled and industrious architects – the ants.

However, the most noticeable feature was the sky, which encircled the whole scene in a boundless blue ocean of unprecedented clarity, so that every detail of the landscape was crystallized in edge sharp focus. This opened up the perspective like a wide-angle lens and one became conscious of Africa's immensity, for here, sitting squarely astride the equator, was a continent big enough to swallow the United States, India, China and the whole of Western Europe with room to spare.

With Jill navigating from a Michelin map we headed southwards, making for Arusha, a town which lies in Tanzania on the lower slopes of Mount Meru. Only twelve miles from Nairobi we turned right, leaving the comfortable tarmac for a rough, weathered surface.

This was our first experience of an African secondary road and I reduced speed to about thirty m.p.h., steering a course between the numerous holes caused by erosion. But very quickly the road deteriorated into a peppering of miniature craters and as these were impossible to avoid I negotiated them at walking pace. After an hour we had covered about ten miles and I made a rapid calculation that at this rate it would take another eighteen hours to reach Arusha and about one year actually behind the wheel to cover our complete route!

Consequently, as it seemed clear that drivers in Africa have no option but to treat their cars harshly, I accelerated. Having settled for a steady thirty-five m.p.h. I found the Cortina took the bumps exceptionally well, but although continuing to slow down for the deepest holes, I still felt rather a brute. Of course, at this early stage in the proceedings it was impossible to know that, in comparison with what awaited us, we were travelling on a particularly fine road; in a few weeks we would become quite oblivious to such obstructions, needing far worse conditions to act as any kind of speed deterrent.

Approaching the small village of Kajiado, Jill drew my attention to a long wake of dust which, way off to our right, was billowing across the Athi Plains. At first it seemed like the vortex of an approaching cyclonic storm although we quickly changed our minds about this and instead decided that it must be nothing more dramatic than a Land Rover.

But on getting close there emerged out of this dust a stream of dull grey shapes and from their curved horns, black faces, white beards and manes we recognized them as wildebeest.[1] The body of this robust antelope tends to slope backwards from the shoulders, and now, with their massive heads carried low, they crossed our path in a peculiar hunched gallop which gave them an oddly grotesque appearance. With every animal violently whisking its tail they stormed south, probably migrating along a known game trail to one of their favourite watering places in the Amboseli Reserve.

The local name for the wildebeest is *nyumbu* and the natives hunt these courageously defensive animals (they have been known to chase off an attacking lion) for their tails, which the Masai tribe use as fly switches.

This was an especially large herd and it was a long wait before the last thundering pair of high-kicking rear legs had passed. At first I had been counting them, but as their numbers seemed never-ending I gave up after a score of fifty. This indeed was a thrilling introduction to wild Africa.

Without seeing any more animals – apart from the distant glimpse of an ostrich, a few gazelle and the unmistakable silhouette of an isolated giraffe – we were soon passing through Kajiado which is fifty miles from Nairobi and situated on a plateau at a height of 5,817 feet.

[1] Also known as *gnu*.

From here we drove into the region named Masailand and when Kajiado was far behind us, we began to see the spear-carrying Masai warriors quite frequently. The men wore knee-length, one-piece toga-like garments made from a strip of orange material dyed in vegetable sap. Fiery red ochre covered their hair which was plaited to the scalp and greased with sheep's fat. Their ear lobes were holed and stretched to incredible proportions, and burdened with enormous hooped ear-rings they sagged almost to the shoulders.

Women dressed themselves similarly but also sported great beaded collars as well as heavy metal arm and leg bands.

Both the sexes are exceptionally tall, arrow-slim, delicately featured and colourful, yet although appearing so formidably primitive, any initial fear of meeting the Masai was quite unnecessary, and as our journey progressed we developed a strong attachment for these flamboyant people. During our travels through Masailand, we moved among them quite freely, actually being invited into their dried cow dung houses which, although interesting, was anything but a comfortable experience.

The border post for leaving Kenya is at Namanga and here we had to fill in exit forms. A customs official sat at a rickety table beneath the shade of a tree and stamped our passports without comment except for asking me if I was Peter Townsend's brother.

Once again the road became excellent, for Tanzania has even surfaced the miles of 'no-man's-land' which separates Kenya from the border at Longido. Here there were entrance forms, a customs check, a police registration and a toll to pay. All these formalities were carried out with typical African geniality, humour and smiles, but nevertheless they took about fifteen minutes and I could well imagine the frustrating delay had there been a queue of traffic in front.

The car toll is 30s and we paid reluctantly, for the normal motorist is only charged 10s and it seems unfair to the visitor that hired cars are classed as *public service transport*. However, it was quite impossible to feel indignant, for every official welcomed us warmly and flashed gleaming white teeth in broad smiles which lit up their dark faces like lightning in a stormy sky. With our toll receipt displayed on the windscreen the barrier was raised and we drove into Tanzania.

Tanzania, covering 342,000 square miles, was formerly the United Republic of Tanganyika but on December 9, 1961, it attained independent status and exactly one year later became a Republic within the Commonwealth. Unfortunately, because of England's failure to resolve the Rhodesian crisis, Tanzania had broken off all diplomatic relations with the United Kingdom. But we soon discovered this was no more deep-seated than a top-level political wrangle and it obviously had no effect on visitors from Britain because the people of Tanzania still exhibited a cheerful, ever-helpful attitude which was already endearing us to the East Africans.

From the border we began to climb. To our left the sheer granite sides of Longido Mountain stood high above us with its sharply curved fang rock protruding from the summit. On the lower slopes were clustered Masai villages with hundreds of cows roaming over harsh terrain to search for grass in the deep ridges, which were formed by gold streaked quartz, white crystalline minerals and glittering mica strata.

Soon, other mountains appeared, but unlike those of Switzerland or similar countries, they conveyed no overpowering, boxed-in feeling, for they rose from the airy plains only as isolated masses. As we climbed higher the hillsides became dry and sandy, being covered with a species of giant, olive-green cactus. Then, just as the road seemed likely to touch the sky we began to descend, twisting downwards towards Arusha. This was the wet side of the mountain range and the sandy slopes miraculously changed into lush, tropical vegetation.

We drove into Arusha at 3.00 p.m. after covering the 172 miles at an average speed of only twenty m.p.h. which was really very slow considering the last ninety miles had been on a first class *transcontinental* road. Having originally planned to arrive for lunch we were therefore famished and thirsty. On future journeys I certainly could not afford to worry so much about the holes; we knew that in the remote south there were several sectors where 500 miles had to be covered in one day on unsurfaced tracks.

For our first few nights in Arusha we were staying at the Mount Meru Game Sanctuary which lies about fifteen miles out of the town. This sanctuary is owned by Dr von Nagy, a prominent authority on wild life. It is an informal establishment for it can accommodate only seven people, and the paying guests mix with the personal friends and acquaintances of the Doctor.

On arrival, we immediately sat down in the garden to a meal of sandwiches, cakes, scones and tea, quickly prepared by the young receptionist. Actually, *garden* is hardly an apt description, for we were overlooking a small natural pool surrounded by forest.

Feeding around the water's edge were six Thomson's gazelles and they stopped grazing to regard us with liquid eyes while their little black tufted tails wagged violently. This party included one ram who was feeding some yards behind the five ewes which formed his harem. All had the same attractive rich, reddish-brown colouring with the inner-side of their rump being snow-white, but whereas the male horns curved gently backwards, the female horns were vertical and exhibited every kind of twisted imperfection.

Generally called Tommies, these gazelle are sweet, confiding creatures and naturally tame but when I cautiously raised the camera they took fright. Holding their legs stiffly and with their heads and tails erect, they ran off in a series of abrupt, bouncing steps as if walking on stilts. This curious gait is known as *strotting* and the manoeuvre is performed by ewes when endeavouring to lure a jackal or some other enemy from where their babies lie hidden. Here, the strotting was a very brief affair and in minutes the gazelles had returned.

There were also numerous exotic birds by the pool and I recognized red-winged turacos, storks, flamingoes and an African broadbill which sat in a tree emitting a reverberative deep trill similar to the British night-jar. But a species I could not identify made a strange noise like a grandfather's clock being wound up. The turacos were especially abundant and these are unique birds not only because their size ranges from dazzling dwarfs to huge specimens over two feet high but also because their colour runs in wet weather, for the red turacin pigment is soluble.

'You must be hungry, so please start to eat,' said the receptionist, offering us sandwiches. 'I've laid a place for Nanny but she probably won't come so there's no sense in waiting . . . Oh, I spoke too soon, here she is.'

I stood up and turned round expecting to see a grey-haired, wrinkly-faced old lady. Instead, I found myself looking at a black, hairy, wrinkly-faced chimpanzee.

For several moments Nanny remained motionless, contemplating the table with angry, suspicious eyes; but, noting there was still plenty of food left, she gave us both a half-pouting smile and swaggered to her seat.

Having quickly confiscated several cream cakes and drunk two cups of tea without spilling a drop, she leaned back from the table, scratched her chest and casually flicked a few crumbs to an Alsatian who had also joined the party. Then, after having assumed an air of complete indifference to us, she eventually gave a distinct sigh and extended a limp, cream-covered arm with the disdainful attitude of a duchess who was being introduced against her will.

We both shook the 'flabbily'-proffered hand but Jill, who adores all monkeys and apes, playfully pulled the long hairs on Nanny's arm. Nanny grinned broadly. Jill pulled the hairs again and if a chimpanzee can be said to laugh, then Nanny most certainly laughed; from that moment she became an inseparable friend.

So, on our first afternoon in Tanzania we were taking tea in the company of a chimpanzee and an Alsatian, while surrounded by gazelles and tropical birds. Then, as a climax to this unusual scene, Africa's highest mountain, Kilimanjaro, suddenly appeared from behind cloud to make one of its rare appearances. Although over fifty miles away it dominated the sanctuary's lakeside view.

'Well, I suppose you would like to see your room,' suggested the receptionist. 'Nanny, you stay here.' This marked the end of our happy tea-party because Nanny did not intend being left behind and she rose from the chair. 'I've told you to stay here'; the warning was repeated with a pointed finger.

As the finger was pointed all hell broke loose. With ear-deafening squeals of temper Nanny defiantly kicked the table and then began stamping on the ground in rage. Alas, the benign duchess had been replaced by a strong-willed fourteen-month-old chimpanzee girl determined to get her own way like a naughty child.

'Stop that at once.' Nanny was given a sharp smack of retribution. At this the squeals were intensified and she sat on the floor banging loudly with hands *and* feet. This was all too much for the Alsatian who rose with bristling hair and growled threateningly: 'Shut up, you spoiled noisy idiot,' he was saying. For this interference he received a hefty chimpanzee punch straight on the nose. With badly-watering eyes he retaliated by making an angry but still tolerantly friendly jump at Nanny who neatly sidestepped, sank her teeth into his tail and then punched him on the nose again as he whipped around to face her.

Of course, Nanny won, and she accompanied us to our room with a boastful walk while the Alsatian walked dejectedly behind, his tail hanging limply.

That evening we sat down to dinner at a long, highly-polished wooden table laid with gleaming cutlery and a silver candelabra. There were no other guests. A white-gloved waiter served us and we ate in the soft light from candles and the flicker of a log fire; as the sanctuary is at an altitude of over 3,000 feet, the nights are chilly. An enormous buffalo head and numerous stuffed birds of prey with outspread wings looked down on us silently. Here indeed was a remarkable atmosphere, for the elegance of a country home was combined so effectively with the incongruities of an African hunting lodge.

Later, Dr von Nagy himself arrived back from safari with a group of friends and until midnight we enjoyed talk about lion, buffalo and other animals. As the evening progressed the stories became even more fantastic and I must have looked incredulous, for Dr von Nagy suddenly leaned towards me.

'You know, things occur out here that are beyond normal comprehension. You may hear of a car being crushed by a giant snake, or of someone who has warded off an attacking lion with his fountain pen. But however unbelievable the facts, rest assured that at some time or another it has almost certainly really happened. And no matter how far you may be from civilization there are always eyes watching. Commit some crime in the deepest jungle, and villages a hundred miles away will know the details in an hour. The jungle telegraph system is absolutely incredible. I tell you, Africa is a strange continent and there's nowhere else like it in the world.'

I believed him, for who could disregard his knowledge and long experience of life in the bush? And on that exciting note we retired to bed.

1. The king of beasts, Ngorongoro, Tanzania.

CHAPTER II

Through Cloud Forest to a Lost World

A famous naturalist – Giraffes with dark curling eyelashes – The elephant bush code – Warty animals – Bubbling water – A six-ton road block – Grace and pace – On foot to the ash cone – Wickedly precipitous slopes – Dodging a rhino – Alpine flora – Whistling monkeys – A lost world – Race against time – Nocturnal babies – No torch and a puncture – Utter darkness – Tremulous screams – Leaving for Ngurdoto.

Mr Desmond Vesey-Fitzgerald is a famous naturalist and a leading figure in Tanzania's animal conservation programme. We had arranged to meet him at his home in the Arusha National Park where he is the Scientific Officer in charge.

This reserve comprises three clearly-defined topographical areas (Ngurdoto Crater, the Momella Lakes and Mount Meru[1]) and its recently acquired national park status has been due almost entirely to the pioneering of Desmond Vesey-Fitzgerald.

(Here I would like to mention that an African national park or a wild life reserve is very different from our usual conception of a park, for it is not *an enclosed area* as defined by the Oxford Dictionary. There are no fences or barriers and the animals are free to wander anywhere, even out of Tanzania and into Kenya or Uganda. A park is therefore only a specified area –

[1] Not to be confused with the Meru Reserve in Kenya which was created by the dedicated work of game warden George Adamson and his wife Joy, who were the foster-parents of that famous lioness Elsa.

2. Lions congregate in families, Serengeti, Tanzania.

although some of them cover as many as 8,000 square miles – where animals are protected, and any killing or trapping of them is a serious criminal offence. Such conservation areas are therefore in no way a superior version of a zoo, and very often, during the migratory periods, more animals can be found *outside* the protected areas than within them.)

We left Dr von Nagy's at 7.00 a.m. Hot sunshine was already dispersing the mist to reveal the steep black outline of Mount Meru.

By now, the hard road surface had deteriorated into no more than a track and our passage left a cloud of dust which swept over the countryside. With wheels spinning in loose sand we negotiated a sharp, twisting hill and then emerged over its crest to find three giraffes straight ahead of us. We stopped immediately, and luckily they did not move.

The giraffe has very acute sight and now they stood watching us with wide eyes, their long, dark, curling lashes and heavy lids giving them an expression of pensive sadness. These were males, reaching the colossal height of what I judged to be about eighteen feet and all three had pale chestnut bodies clearly marked with a jagged pattern like crazy paving, although below the knees they became spotted.

What lovely animals they are! Even more beautiful than a Walt Disney creation. The sheer thrill of seeing them for the very first time in their natural habitat, and at such close quarters, inexplicably brought one to a point of exaltation.

Slowly, they began walking away with a gait similar to that of the camel whose left and then right legs move together. Next, with their tails triumphantly curved upwards they broke into a run and pranced through the bush in an unusual 'slow motion' gallop whereby the hind limbs gracefully came well ahead of the forelimbs. Each effortless leap took them around fifteen feet forward and their necks rocked to and fro like ships' masts on a rough sea. Once at a safe distance they paused and turned towards us again, staring curiously with a demure maiden aunt appearance. I felt that a flower-decked hat would be in keeping.

Although a giraffe's skyscraper neck contains no more vertebrae (seven) than the human neck, the supply of blood to such a height requires a pressure thirteen times that of man's. It is believed that their jagged white coat pattern indicates the interlocking of a special system of blood vessels though this remains unproved; but, whatever the anatomical reason, these features are the giraffe's trade mark.

They are also extraordinarily gentle creatures, rarely being vicious even when wounded, yet if attacked by a lion, their gigantic hooves can deliver one-ton sledge hammer blows that are more than sufficient to crumple up the aggressor. However, when two bulls have the inevitable argument over a female they never use these powerful hooves and instead endeavour to knock each other with their horny protuberances, but such disputes are generally so refined that, from a distance, the weaving, probing and evading is almost indistinguishable from love play.

With the giraffes still watching us coyly we moved off again. After a few more miles the track narrowed until bushes touched the car on each side, making it impossible to avoid innumerable sharp-edged rocks which jutted from the ground. Squirrels continually darted across our track, and troops of baboons peered from the long grass.

Then we startled a pair of warthogs. Unlike the giraffes they were not exactly majestic and definitely not beautiful. Their faces had enormous warty lumps and, just above massive scimitar tusks, stiff whitish bristles sprouted backwards from the cheeks. The warts, the tusks and the bristles gave the impression of fat, white-moustached grumpy old colonels. Trotting away in alarm, their slender tails were held erect with the extreme tip wobbling.

Warthogs are the commonest wild pig of East Africa and although courageous when attacked they are never as aggressive as the smaller bush pigs or the giant forest hog which the natives fear more than a buffalo.

When we reached Mr Vesey-Fitzgerald's house the mist had gone completely and the forest-covered slopes of Mount Meru were perfectly clear.

Desmond – as he asked us to call him – was a mature man with grey hair and a complexion heavily tanned from long exposure to the tropical sun and wind. His body was strong and rugged for most of his life had been spent out of doors studying the animals of Africa. He wore a pair of khaki shorts, a jungle green open-necked jacket, tough climbing boots which reached almost to the knee and a waterproof watch set into a thick, self made, black leather strap for additional protection against impacts. Here was a man obviously used to coping with unexpected danger and hard conditions. If we were going to meet a furious rhino then I could think of no one more reassuring to have at my side.

Leaving the Cortina we crammed into the front seat of Desmond's Land Rover; the back was full of building materials for the construction of a mountain shelter. He was taking us to Meru, a new sector of the Arusha Park which was being developed for its scenic attractions rather than its wild life. Although not yet open to the public, a track had been cleared up to a height of almost 8,000 feet and this followed an elephant path, for the elephants' efficiency at discovering the easiest gradients over hilly terrain led many of the pioneer roads to be constructed over their ancient, well-trampled routes. From where this track ended we would begin our climb to Meru's volcanic ash cone in the crater centre; only a handful of people had been there before us.

'If this is an elephant trail, what happens if we meet one?' asked Jill, who had obviously noticed the preponderance of broken trees and the huge, roundly oval tracks which clearly showed the wavy, rugged skin of a sole imprinted upon them.

'Good gracious, someone should have explained the elephant's bush code to you before this,' replied Desmond. 'Unless you understand their code it's just as dangerous as being in a foreign country and not knowing whether to drive on the left or the right.' He sounded genuinely horrified.

'If motorists come face to face with an elephant they must stop – with the engine still running of course. Normally, the elephant will first show interest or displeasure by bringing his great ears forward as a prelude to ambling away. But if in a cantankerous mood, he'll raise his trunk, give blood curdling screams and madly flap his ears like a pair of wings. They're often fifteen feet across so you can almost feel the draught. He certainly leaves you in no doubt as to his annoyance. That's when . . .'

At this point in the conversation our Land Rover hit some particularly vicious ruts which brought deafening crashes from the back and sent us all bouncing to the roof. It was several minutes before we gained complete equilibrium and then Desmond continued as if there had been no interruption.

'That's when you reverse smartly and by doing this it shows old Jumbo respect and there's no need for him to prove he's the master. Even in the bush, elephants adopt a very similar procedure among themselves. Naturally, if you try to bulldoze a way past, or fail to reverse, then you're just inviting him to knock the car around for a bit of fun.'

Once again Desmond was interrupted. This time we were descending a steep, soft earth ramp to cross a fast-flowing river. Although not wide, the bubbling water was about two feet deep and momentarily our wheels lost their grip, sinking into the muddy bed. At once the current twisted us sideways, but luckily the four-wheel drive gained a hold and with spasmodic crab-like movements we emerged on to the opposite bank.

After this, rain forest closed us beneath a canopy of leaves and as the track snaked upwards the air became cooler, for with every 1,000 feet rise in elevation there is a corresponding temperature drop of three degrees Fahrenheit. Occasionally, the sun managed to cast fugitive rays into this grey penumbra and, like a searchlight beam it would penetrate the tangled luxuriance of vegetable life to reveal orchids and other flowers glowing vividly against the darker foliage.

As if fate had overheard our previous conversation, we rounded the next bend to find an elephant blocking the way. This was a monstrous old bull which would weigh somewhere around six tons. African elephants are not only taller and bulkier than the Indian or Asiatic species, but their tusks are very much heavier.

Jumbo now watched us with eyes that slowly opened and closed in a methodical rhythm. Characteristically, his face appeared fixed in a permanent smile as if laughing inwardly at some obscure witticism, while the overall expression was one of patience and wise understanding. It seemed hard to believe that an elephant can be extremely vicious and by nature is a temperamental, moody, unpredictable animal, although an unmolested bull is usually more placid than the nervous, touchy females.

On this occasion we were not destined to see any display of ill temper and without even twitching an ear this particular elephant unhurriedly began ambling into the forest with typically delicate steps as if walking on thin ice. Almost silently he disappeared amidst the dense vegetation and our last impression was of his lolloping skin which hung in deep folds like an outsized suit. Perhaps this very precise, unruffled attitude accounts for their longevity of well over seventy years.

Of course, the mere size of elephants give a deceptive slowness to every movement they make, yet their average walking pace is in fact estimated at four m.p.h. and this can be increased to ten m.p.h. under alarm. Due to a great length of stride, distances are covered quickly even when apparently shuffling along, while a charging or badly scared elephant has been known to achieve short speed bursts of up to thirty m.p.h.

How such enormous creatures can tread so gracefully is quite remarkable. Indeed, naturalists say that an elephant has adopted such an effortless gait that their feet exert relatively little pressure on the ground and consequently they produce far less noise than many of the smaller animals. Yet, adult elephants are so powerful that they have no natural enemies and are even perfectly able to kill the tough, amphibious crocodiles which sometimes undertake overland travels between one pool or river to another.

Some elephant encounters are far more dramatic and Desmond told us that, only a few days earlier, six British nurses had met one while driving themselves to Ngurdoto Crater. Unfortunately, when he began the ear-flapping charade they panicked, jumped out of the car and ran.

'Well, poor old Jumbo didn't know what to do with screaming nurses racing everywhere,' said Desmond. 'After all, they were breaking the rules, so he became agitated, swung his trunk around and accidentally knocked two of the girls over as he was wandering off. Neither of them were hurt but all six arrived back at my house in a state of hysteria and told me an elephant had just attacked them. That's a perfect example of how ridiculous stories begin,' he snorted.

Nodding vaguely, I failed to admit that if an elephant had knocked *me* down, then whether or not it was my own fault for playing the wrong game, I would still feel victimized!

Our steep ascent continued with the Land Rover in first gear and its engine roaring loudly. Then, after about forty minutes we broke out of the forest and emerged on to a small, sunny plateau where the road ended. From here there is a magnificent view of Africa's grandeur.

Far below us lay the Momella Lakes, each of them having different shades of water which varied from a forget-me-not blue to an emerald green. The sun, catching on their numerous arms, reflected into the deep clefts of distant volcanoes whose lava black peaks, being embedded with mica crystals glittered like tiaras. The mightiest of these extinct volcanoes was forty-four miles away – Kilimanjaro, whose eternal snow-capped dome punctured the clouds and soared impressively to 19,340 feet, its scintillating ice walls a kaleidoscope of colour.

To the extreme west, an isolated storm was sweeping through one of the valleys and rainbow tentacles were falling from the black sky like carnival streamers. Yet nowhere in the panorama was there any sign of human life or habitation, and the variegated pattern of clouds moving across the rolling plains stamped the final touch of vast splendour on this wilderness.

Immediately above us was Mount Meru; geologists say this would have been even higher than Kilimanjaro if its crater had not ruptured in a series of violent explosions about a quarter-million years ago. Meru's summit now reaches 14,979 feet and the whole mountain is made particularly imposing by a wicked rock face which forms an overpowering backdrop to the ash cone as it rises perpendicularly for all of 7,000 feet.

With Desmond leading, we set off from the plateau in single file, climbing steeply, pushing between thicket and scrambling over volcanic debris which had once obliterated this countryside. After thousands of years nothing has changed here. The numerous boulders that still lie scattered throughout the vegetation might easily have cooled but yesterday; these slopes of Meru are ageless.

Animal droppings were everywhere and Jill immediately asked her inevitable question.

'Will it be dangerous if we come across a buffalo?' Due to the altitude and exertion her voice sounded breathless and I think Desmond mistook it for fright, so he gave an apt but nonchalant answer.

'Well, it's no more dangerous than being in London. After all, a bus can do a lot of damage if you walk in front of it.' He laughed. 'Seriously though, animals are usually only vicious when they are wounded. Of course if you take them by surprise then they'll charge by instinct. A buffalo is particularly dangerous because he charges with the head well up and only lowers it to toss at the last moment. But the poor old rhino comes at you with his head down and providing there's time he can generally be dodged – then he goes thundering past and on into the distance like a runaway train.'

While talking, Desmond had sat on a rock and this gave us both a much needed breather; being unaccustomed to high altitude the climb was proving to be rather strenuous. However, this respite provided an opportunity to admire the flora. On these cool slopes there are many varieties of alpine flowers as well as shrubs and plants indigenous to Europe. The damp misty atmosphere also encourages a heavy growth of mosses and ferns, their blanket of sombre green enlightened by splashes of bright red balsam flowers which are often called 'touch-me-nots' because the ripe seed pods burst on contact with the hand.

Eventually the trail flattened out until we were walking easily over gently undulating hills which were covered by thin rain forest. There was nothing dark or jungly about these woods. They were airy, fresh, lit by filtered sunshine and had an undergrowth of young grass instead of tangled creepers.

A sudden crashing noise and a series of shrill whistles from one of the trees startled us and we looked upwards. A troop of the rare colobus monkeys were leaping sure-footedly from branch to branch, their silver tails spread like fans and standing out in glaring contrast against jet black bodies.

The colobus has a serious disposition and rarely frolics around, preferring to remain concealed in high altitude forest retreats where few visitors penetrate.

Unfortunately, colobus fur is the traditional head-dress of the Chagga tribe and consequently this species has been hunted ruthlessly. It is a great pity. Apart from being the most beautiful of monkeys they are also quite harmless, living almost entirely on leaves and vegetable matter. Although mainly silent during the day – except when disturbed – their early morning and late evening soprano chorus swells in unison, its hymnal quality echoing through the mist-girded forests.

Within seconds the whole troop had disappeared into the deeper foliage and all was quiet again as they lay camouflaged in safe hideouts, no doubt watching our departures closely. But even such a brief appearance was lucky for us, because they are shy and uncommon.

Further on, the vegetation grew sparser until finally the scene opened up to reveal a small, circular lake with Meru's razor-sharp escarpment – formidably powerful at such close quarters – reflected in the motionless green water. Nothing moved on this landscape; not a blade of grass rustled; not a tree stirred.

None of us dared to break the spell of this solitude and we continued our way in absolute silence. Then, after skirting the water and scaling one more long hill there lay before us an extensive area of tableland encircled by vertical black cliffs. Here was the centre of Meru's ash-cone crater.

This was truly a lost world. An oddly colourless landscape, bathed in cold sunshine and belonging to another dimension of time. Even the trees exuded an eeriness. Centuries ago they had perished in a fire, but now their lifeless frames were enveloped in lichens.

Commonly termed *old man's beard*, these pale yellowish-grey lichens, a strange combination of fungus and algae similar to seaweed, hung in thick, cellular strands like bizarre Christmas decorations. The general lack of any dominant colour gave the impression of looking through steamed glasses, and this effect possibly accounts for it being called a *cloud forest*. Yet the ghostly appearance was attractive. There was no underlying atmosphere of anything unpleasant or supernatural, only the fascination of a world untrodden by man for hundreds and thousands of years.

On the northern and eastern sides, the land fell away into the valley far below, and with the aid of powerful binoculars we sat watching buffalo, rhino and elephant, which to the naked eye were no bigger than ants. We also picnicked here and, finding it quite impossible to ignore the crater's enchantment, we did not move until the sun had disappeared behind Meru's twin peaks. Almost at once the air turned chilly. Clouds appeared, tumbling down the precipices and surrounding the ash cone in a white halo. Fog patches had already formed and our clothes suddenly felt damp.

'We had better hurry, otherwise there'll be an uncomfortable overnight stay,' said Desmond.

A second warning was unnecessary and we began the return journey at a fast pace. In the rain forest it was very murky but Desmond forged ahead without any hesitation. 'If we followed that path there'd be a nice big cliff to fall over,' he informed us casually, pointing with his pipe.

But once out of the forest and away from the crater, the sun was still shining and darkness fell only after we had made our final descent in the Land Rover.

Before leaving we arranged to meet Desmond the following day at Ngurdoto Crater, and after he had drawn us a sketchy map we waved goodbye and drove off into inky blackness.

The eyes of numerous animals reflected constantly in our headlights, blazing points of red, green and even yellow. Some were round, others oval or merely slits. The smallest of them glowed from the thorn bushes like cigarettes burning and as these were obviously moving among the trees they probably belonged to the nocturnal bush babies.

A few miles after leaving the narrowest portion of track our rear tyre went flat. Foolishly I had left the torch at Dr von Nagy's, so I found the jack and got it in position only after a great deal of fumbling. By this time our presence had attracted company. Two elephants were coming along the road towards us so I jumped into the driving seat and switched off the headlights; like the hippopotamus, an elephant will often attack the lights of a vehicle.

Sitting there, waiting, made the blackness seem even more intense. Not a star was visible.

'Do you think they're still around,' Jill whispered eventually.

To answer her question I flashed the lights briefly and we saw them quite clearly, both standing some way off, but still too close for comfort. However, my eyes were now becoming accustomed to the darkness, and rather than wait any longer I decided to change the wheel without using the lights again. Apart from temporarily losing a nut and being unnerved by the tremulous scream of a tree hyrax almost directly into my ear, the operation went smoothly and quickly.

There were no further incidents before we arrived back at the Mount Meru Game Sanctuary, but then, just as we entered the driveway, our off-side *front* tyre went flat. It was rather late in the evening to worry about getting tubes patched, so before retiring I jacked up the car to prevent the tyre wall from being crushed.

The next morning I drove at walking pace to a nearby garage and after both punctures had been repaired we hurried off to meet Desmond at Ngurdoto Crater as arranged.

CHAPTER III

Wild Life in Peril

A great terrestrial upheaval – Unique conditions – Financial greed and destruction – Conservation or extermination – Figures speak louder than words – An eternal struggle – Man versus nature – Filth, pollution, nightmares – Sagacious people – The fools – Local complications – Date with a rhino – An irreplaceable fairyland – Two men who will never be forgotten.

THE origin of Ngurdoto Crater dates back fifteen million years, from when the Great Rift Valley was formed after the earth's crust had split asunder from the mouth of the Zambesi to the Turkish border. This great terrestrial upheaval opened a small volcanic vent at the eastern foot of Mount Meru, and continual explosions within the same area eventually caused the formation of Ngurdoto which now covers an area of almost four square miles.

Although not as spectacular as Meru, it is unique because the crater floor has no access and the countless animal herds which live there have never been *disturbed* by man, let alone hunted.

As the crater offers excellent grazing, permanent water and convenient mud wallows, buffaloes are always prevalent and we looked down upon a herd which must have been at least 300 strong. There were also many warthogs, rhinos, elephants and baboons, which appeared to be spread over the grassland like ornaments carved in various shades of black, grey, brown and russet woods.

'Isn't that a beautiful sight – they haven't any idea what man even smells like,' said Desmond. 'But believe it or not I'm having constant battles with the local tourist board who want to build a road down into the crater and have little Volkswagen buses tearing around for sightseeing tours. Can you imagine that!' He gave a deep sigh. 'Of course, I want the tourists to come but only if they can appreciate Ngurdoto as it is. There's no other conservation area in the world where the animals are so natural, simply because humans have never intruded upon their existence. Besides, this particular crater is very much too small for tours. If animals are harassed in such a restricted area they either die, migrate, or become no more than zoo creatures.'

I could understand his concern. The economic pressures on him must be enormous. Until he allows the crater to become a Volkswagen merry-go-round, then local authorities and tour operators are showing complete disinterest. But the Arusha National Park is largely dependent upon the revenue from visitors, and without these funds all the pioneering work in this area could easily go to waste.

Thanks to men like Desmond – Europeans and Africans – Tanzania, together with Uganda, has the biggest concentration of wild life to be found anywhere. Even so, many reptiles, birds and mammals have disappeared in the last half-century, and like the dinosaur, pterosaur, dinornis, pterodactyl and scores of others, they will make their appearance only in books as an artist's impression.

Just three years ago, in Uganda, the mountain gorilla was officially classified as extinct. These were King Kong size monsters standing ten feet or more high, but they were quite timid, and when man began to penetrate their mountain jungles on sightseeing tours they became rapidly depleted. Very few photographic records of them exist. They were harmless creatures, keeping away from civilization, and in spite of their immense size lived mainly on roots and berries.

Comparatively few animals breed successfully or prolifically in zoos, and so serious was this problem of mass extinction that it was once doubtful if our grandchildren would be able to see a common lion or a leopard. Even the elephants faced extermination when, between the years 1888 and 1908, their numbers were reduced from three million to 280,000.

Certainly there is no justification for destroying the unique attraction of Ngurdoto. Admittedly a few local businessmen could increase their bank balance over a short period, but I am sure that if the tourists had any say in the matter they would want this crater to be left unspoiled. Although there are always people who are willing to destroy anything for personal gain, I have every confidence in Desmond and his supporters to win the battle.

This confidence was strengthened even further when another Land Rover arrived. It was driven by 'Mike', a Britisher and one of the Park's Educational Officers. He was thin and wiry with short, cropped hair. Appreciably younger than Desmond his face was just as tanned and he shook hands with an iron grip.

We sat talking for a long time while Mike trotted out facts and figures like an encyclopedia. Certainly it was surprising to learn that, because the Africans tend to live in densely populated areas, very few of them have seen any of their animals, and when films are shown they get very excited and invariably shout 'lion' at the hyena and 'elephant' if a tusky warthog makes an appearance. Only a tenth of them ever visit the parks.

When Tanzania became independent, the Africans – even those who were educated – thought the land devoted to National Parks could be put to better use, and that animal viewing was some quirk of the white man.

But wild life experts such as Mike immediately emphasized how much foreign currency the parks would bring in to the country, and then organized a mass instructional programme.

Fortunately, East Africa is an extremely intelligent and responsible society. It was not long before the Governments of Tanzania, Uganda and Kenya recognized the necessity for conservation, not only as the means of building a successful tourist market but also because they realized the animals were an important part of their heritage and must be preserved for posterity.

The need for maintaining the national parks has therefore become fully recognized, and many new ones have been created. But there is a formidable task ahead and, if the perpetuation of wild life is to be assured, the population must be educated quickly.

Even now, there are still a few loud voices of protest, and

many of East Africa's beautiful birds and animals continue to be slaughtered or hunted as sport. Poachers still succeed in making a vast number of illicit kills and they roam the parks using wire snares, cage traps and poisoned arrows. The areas concerned are so vast that adequate supervision is difficult, and although heavy penalties are levied on poachers if caught, there is a tremendous black market in skins, feathers, ivory and rhino horn.

In certain oriental countries, powdered rhino horn is regarded as a powerful aphrodisiac and it can be sold for almost 10s per ounce, four to six times as much as ivory. Consequently, poachers annually reduce the rhino herds by one-fifth and their birth rate cannot compensate for such a heavy loss. White rhinos have suffered particularly and the rarity of this animal can be judged by its catalogued price; they cost a zoo around £5,000 compared with less than £1,000 some years ago. There were once thirty-three species of rhino, while today only five exist.

The craze to wear clothes decorated with silky fur, or made entirely from attractively marked pelts, is yet another financially rewarding market which causes the unlawful slaughter of a further half million animals yearly. From Tanzania alone, 50,000 colobus monkeys are poached and there are now little more than 500 cheetahs left in the whole of East Africa.

The shortage of grazing land is another reason for the rapid dwindling of wild life. Where not so long ago the Serengeti claimed a million head of big game, recent surveys have shocked conservationists by showing that the total concentration has been reduced to perhaps less than 400,000. And around Lake Bangweolo in Central Africa, herds of waterbuck have decreased from 150,000 to 10,000. Such figures can be quoted endlessly and they speak stronger than any words.

However, the problem does not have its roots in Tanzania for this situation is no more than yet another variation of man's eternal struggle for scientific advancement, economic superiority and power. Commercial expansion, organized industrial development and monetary gain are nature's most bitter enemy. The constant battle is not confined to Africa.

In the blue Pacific Ocean hundreds of square miles are

contaminated with radioactivity and classed as restricted danger zones. The tropical islands of Mururoa and Hao have been erased from the earth by French nuclear tests. Italy has a higher index of industrial expansion than any other European country and the once famous Riviera coastlines of Liguria and Tirreno are now a nightmare of chemical plants, natural gas refineries and phosphate quarries.

Over much of the world, rivers and lakes are polluted with the filth of what we call progress; their water no longer supports any fish life. Forests are stripped and pastureland lies ruined. Construction engineers have blasted their way across the landscape with dynamite, scarring the countryside with an endless succession of roads, towns, factories and airports. Nature has been pushed farther and farther away and suffocated by brick, concrete and bitumen, while at every opportunity the wild life that remains has been persecuted.

But out of this turmoil emerged prudent men and women who realized that humanity was racing along the track of self destruction. They realized that one day there would be no trees, no flowers, no birds, no animals, no fresh air to breathe. They realized that the earth may become barren and our rivers and lakes revert to smelling swamp land. Our whole existence was in peril and after only another century the great outdoors and its stimulating magnificence may be gone for ever. And with the balance of nature irrevocably altered what pests and strange diseases would befall everyone?

These sagacious people then raised protests. They formed official groups. They founded societies. But all had the same aim – to gain support for protecting nature. Here indeed were voices shouting in a wilderness which was ruled by the stubborn, the fools and the deaf. But the voices grew in number and among them were men like Desmond and Mike. The *animal conservationists* had been born.

As an educational officer, Mike is now utilizing every modern aid to teach the people about conservation. Mobile film units tour remote areas, colourful posters are displayed in even small villages, guided tours of the parks are arranged and free accommodation is provided. The work is carried out with an urgent conviction. But as Mike said, 'It is absolutely vital that

3. **Warthogs take fright, Uganda.**

the next generation – including all those who will neither read nor write – are taught to protect instead of to kill.'

It is a mammoth undertaking. Adequate funds are needed and in every nation more voices must be raised to support conservation; this affects all civilization and is part of a complex effort to exist. In Africa, the operation is even further complicated by local factors. The Masai tribe are short of land and water for their cattle, and quite naturally they resent the dik-dik, zebra, gazelle, bushbuck and all the other creatures who encroach upon their pastures. Therefore, outside the reserve areas, animals continue to be exterminated.

Certainly I can sympathize with the Masai because they honestly think that wild life is a threat to their survival. Of course it is a fallacy, but the Masai are convinced of its truth and as these people once ruled nearly every square mile of East Africa their feelings, even if misguided, are justifiably strong.

Yet, after I came to know the Masai better, I realized this deadlock was not insurmountable, and Sir Julian Huxley's wise suggestion of making certain regions *Masai Parks* is one of the many possible solutions. Giving the Masai a share of the profits yielded by tourism would convert them from enemies of wild life into responsible conservationists.

'Well, I really must away,' said Mike, looking at his watch. 'As you can imagine, we all have our hands full. It's been a great pleasure to meet you both and I do hope we can get together again.' He turned to Desmond – 'By the way, thanks for keeping your rhinos on a lead this morning.'

'Yes, we had quite a shindy that morning,' replied Desmond. 'Have the cuts all healed?'

'What happened?' Jill asked.

'Last Tuesday we were standing just where you are now when a rhino burst out of those bushes.'

'I thought you said that animals only attack when they've been wounded,' Jill remarked with feigned indignation, but nevertheless surreptitiously glancing over her shoulder.

Desmond laughed. 'Now who said anything about being attacked? The rhino merely came out of the bushes and found us unexpectedly in his way. This frightened him, so he charged

4. Hippos are rarely seen out of water. Along the Nile, Uganda.

us and we both had to climb a tree.' He pointed to some frail, spiky-leafed saplings that looked hardly strong enough to take the weight of a spider's web.

'As always with a rhino it was a case of out of sight and out of mind, but then he began to graze. By this time the branches of both our trees were sagging a bit . . .'

'You can say that again,' cut in Mike.

'. . . so I threw a box of matches,' continued Desmond. 'When the rhino charged off to where it landed, Mike jumped down, scooped up some stones and promptly whipped up a larger tree. Then he used a stone to do the same for me and the whole operation was repeated several times until eventually we got to safety. Quite an interesting morning, really,' he concluded with all seriousness.

After watching Mike's dust trail disappear Desmond also had to leave, and although we spent another week exploring the Meru and Ngurdoto area, our paths never again crossed.

Neither Desmond nor Mike will be forgotten. They live alone in the bush, working with others to preserve an irreplaceable fairyland for everyone to enjoy. While such men deserve our respect and appreciation, even more they deserve our support. And we can all give this support by taking a genuine and sympathetic interest in wild life. An understanding of nature will help to save the world's few remaining animals from being decimated to extinction like so many other splendid creatures before them.

CHAPTER IV

The Illusive Mountain

The administrative centre – A fascinating spectacle – Three extra wheels – Scientists scoff at explorers – Immortalized by Hemingway – Fatal charm – Indian filth – Misrepresentation – Cockroaches, ants and weevil beatles – The Marangu – Mountain sickness – An attempt to 18,635 feet – Snow intervenes – Tragic death – The real struggle from Kibo peak – An unusual photographic opportunity.

BEFORE leaving the Meru area we stayed for several nights in Arusha, the capital and administrative centre of the entire Northern Province. With around 10,000 inhabitants it is a small, modern, attractive town. It is overlooked by villages built precariously on the mountain slopes, but so well camouflaged amongst the coffee and banana plantations that at first sight they may remain undetected.

Wandering around Arusha's shops became an enthralling pastime, not for the goods displayed but for observing such an interesting, colourful mixture of incongruous people.

An African businessman dressed in a lightweight suit with collar and tie stands next to a WaArusha warrior who wears a loin cloth and carries a spear. Towering behind them is a Masai with red dyed hair, a toga, the inevitable spear and brass rings shining in his elongated ear lobes. There is a dusky beauty showing off a smart miniskirt and a WaIraq woman wearing a strange black garment decorated with coloured beads in mysterious patterns. They are all peering through the same shop window at refrigerators! And what a fascinating spectacle this crowd of people make.

Staying at the New Arusha Hotel our room gave us airy comfort and the food here was suitable for a gourmet. It is a pity that at the present time Arusha has no international airport, but there are regular DC-3 flights from Nairobi. If anyone is contemplating a visit to East Africa, Arusha is undoubtedly the perfect location for a base or a 'kicking-off' point; it is near to some of Tanzania's most scenic wild life areas and ideally positioned on the Great North Road at a central spot lying half-way between Cape Town and Cairo.

From Arusha our next destination was Moshi. On the morning of departure I found both front tyres flat. With punctures occurring so frequently I began to worry about conditions out in the bush when we were hundreds of miles from anywhere.

As there was a UTC office in the town I decided to ask them for six spare tubes and some more patches. After discussing our route with them they insisted that in addition to the tubes I should also take an extra two wheels as well as the spare one which is standard equipment. (Wheel rims often get so badly damaged that changing a burst tyre is impossible.)

At the same time we were refitted completely with new Firestone tyres, which apparently have proved themselves to be as tough and reliable as the Ford Cortina for withstanding the dreadful road conditions that often ruin tyres of lesser quality after just a thousand miles. In fact, I had never used Firestone products before but naturally local advice and experience was unquestionable. We were also told to carry food and water supplies whenever we undertook long hauls.

As the road to Moshi took us past the Mount Meru Game Sanctuary it was impossible to resist bidding Nanny goodbye. Then, having shaken her shrivelled little hand for the last time we set off towards snow-capped mountains on yet another fast, smooth highway.

Every mile we travelled brought Kilimanjaro closer; Moshi is situated at the foot of this mountain, which is yet another extinct volcano. The Masai call Kilimanjaro *Oldoinyo Oibor* (The White God) and its Kibo peak *Ngaje Nga* (The House of God.) This is the highest point on the African continent, and although only three degrees south of the equator the snow-cap is a permanent feature.

More often than not, the whole of Kilimanjaro is obscured by cloud, and this is why it remained undiscovered for so many years. In fact, when explorers first began to talk about sighting a great snow-covered mountain on the equator they were scoffed at, and later expeditions which never found Kilimanjaro increased the general belief that it must have been a mirage. Then, during 1848, Rebmann pinpointed its exact position but, when he made official reports, members of the London Royal Geographical Society (who had never been to Africa) told him categorically that he was mistaken! Eventually, the existence of the mountain became a subject of constant debate; it can be compared with the *Abominable Snowman* controversy.

But the reputation of Kilimanjaro's beauty has now spread through every continent. It has been immortalized by Hemingway. And today, this mountain is undoubtedly one of the world's most famous natural features.

On this particular morning, no explorer could have thought Kilimanjaro was an optical illusion. The sky was azure and this illusive mountain dominated the entire northern landscape, rising from the ground with the delicacy of a ballerina. Unlike Longido or Meru there was nothing harsh or rugged about its contours. They were gentle, softly rounded and intimate, like the body of a beautiful woman veiled in wisps of violet chiffon. But this was no artist's colour to be captured on canvas. It was unrefined and exciting to the senses. A colour which had the mysterious quality of exuding sweetness and danger, while the snowy head-dress, seemingly fashioned in bridal silk dazzled the eyes with its virgin purity.

Here was the spirit of a real mountain goddess, but as with the legendary sirens her invitation was not to be accepted lightly; every year men succumb to Kilimanjaro's fatal charms.

Due to a large Indian population, the town of Moshi is unattractive and dirty.

Only one of its four hotels had vacancies; its claim to be 'The Modern Luxury Hotel of Tanzania', together with a lavish brochure, is one of the most flagrant examples of misrepresentation that I have ever found.

Our antiquated room had no fan and was oven-hot. The walls were filthy and covered with smears of blood where insects had been killed. Ants nested in the bath over flow, and a constant stream of them paraded across the floor. The state of the toilet seat is best left undescribed, and in three of the newspaper-lined dressing table drawers there were very large and active cockroaches. More cockroaches were in the grubby wardrobe; when I complained, the surly Indian receptionist simply told me to ask the room boy for some DDT! The finale came at breakfast when Jill poured milk on to her Rice Krispies and immediately dozens of weevil beatles began to evacuate. 'They're quite harmless,' the waiter informed us.

We therefore stayed at the Livingstone for only one night before moving to the Marangu, a small but extremely congenial hotel twenty-five miles away on higher ground. Many people begin their ascent of Kilimanjaro from this hotel, and those unskilled in mountaineering can choose a comparatively easy route, although even this one requires great physical endurance over five days. Because of the altitude climbers often experience mountain sickness, and this alone can make the summit inaccessible.

Neither Jill nor myself are very happy at great heights, but we decided to make an attempt as far as Gillman's Point at 18,635 feet. However, fate intervened and not long after the coffee, maize and banana plantations had degenerated into grass and steep moorland, cold mist was beginning to form. By the time we reached Mandara Hut at 9,000 feet our energy had been sapped by thick cloud and light sleet showers which sent unpleasant shivers through our clammy bodies. Our guide advised an immediate return and his fears were fully justified; during the night tremendous storms brought the snow line down to an unprecedented 10,500 feet.

Even experienced climbers had been forced back long before reaching Gillman's Point. Only one life was lost, but this was particularly tragic because the climber concerned, apart from being with a competent party was a young man apparently in excellent health, yet he died from exertion during the bitter cold at around 19,000 feet.

Naturally, there is less possibility of such a disaster happening on the easier route from Marangu. Even so, bad weather and mountain sickness often delays parties in the Kibo Hut which has been constructed by the local Mountain Club at 15,520 feet. After Kibo the ascent is quite hard but it is from Gillman's Point where the real struggle begins and to reach the summit means a very tough one-and-a-quarter mile trek around the crater followed by a final 750-foot climb in deep snow.

But *we* were destined never to reach even Gillman's Point and in spite of sunshine on the Marangu Hotel, Kilimanjaro continued to sulk, making us realize how lucky we had been on the previous days when this famous mountain was revealed in such splendid glory and I had seized the opportunity to take photographs. When we left the area finally, the *White God* still remained obscured by cloud.

Our new destination was Taveta.

CHAPTER V

Meet Pat Hemingway

Introduced by proxy – Taveta District Court – Poaching – An infectious laugh – The United Nations – Beer instead of water – The urge to write – A personal insight – Green students – Canned elephant – Life with a purpose – The judge arrives – Translation difficulties – Detailed cross-examination – No witnesses called – Police contradictions – Case dismissed.

TAVETA is in Kenya, a short distance from the Tanzanian border. We were going there to visit Pat Hemingway, son of Ernest Hemingway; the invitation had been arranged by a mutual friend in Nairobi.

Of course, I was particularly nervous about this meeting, knowing that I had been introduced by proxy as author of the *Quest* series. It must have sounded rather trite to someone whose father had won the Nobel Prize for literature.

We had been told to make direct for the Taveta District Office Court and on arrival we were surprised to find large groups of people everywhere, numerous police and many park rangers.

Immediately we left the car, a short, rugged, stockily built man with brown hair came over to us.

'You must be Derek and Jill Townsend. Well, I'm Pat and you're most welcome.' He shook our hands. 'As you can see I've asked you to come at an interesting time. We have our first court case. That European over there' – he nodded – 'was found shooting guinea-fowl in Tsavo National Park. Not that guinea-fowl are so valuable, but it's the principle which lies at stake. Think how the London police would feel if they caught some educated fellow using a rifle to poach the ducks in St James's Park.' He laughed and his weather-beaten face at once crinkled into familiar lines of natural merriment. Suddenly both of us were laughing. Pat's humour was infectious.

'Anyway, the judge won't be here for another hour so let's sit in the shade.'

We all squatted beneath a tree and were joined by Frank Poppleton, Tuffy Marshall (the Senior Park Warden of Tsavo), Hans Reinwalf (from the German Development Aid of the Federal Republic) and Akarbar Din of the Ugandan Government. Pat and Frank are both working for the Food and Agricultural Organization which is part of the United Nations Development Programme. This is a Government body comprising people from the three East African countries (Kenya, Tanzania, Uganda) and their specific project is the training of Africans to become Game and Park Wardens.

Pat is an instructor at the College of African Wild Life. 'I've been doing this job since 1964,' he said. 'I came out to Africa in the days of colonialism and very little has really changed. Not even the water.' He gave an impish, naughty-boy look and his eyes twinkled mischievously as he nudged Frank. 'Tell them what we told that health officer who came out to examine our first water supply.'

'Yes, that was rather amusing,' said Frank. 'This health officer came out and asked what precautions we were taking with the water so I gave him the story that having boiled it and added purifying chemicals we finally used a filter. Well, this seemed to make him very happy, but then Pat spoiled everything by saying, 'And after filtering, just to be on the safe side we drink beer instead.'

'We still do,' added Pat, and once again we were all laughing.

As the judge had still not arrived by noon, Pat suggested a quick lunch at the Ranger Post he was using as a base. This was almost thirty miles away and shows how distance means very little out in the bush, for by 2.00 p.m. we were due back at the Court House.

Several other groups were joining us, so we merely tagged on behind the small convoy of Toyoto jeeps, following their dusty wake into Tsavo Park which covers 8,000 square miles of Africa.

Over a delicious stew Pat talked continually, and although I never asked him anything about his father he did volunteer the belief that all authors should experience a desperate need to write, an urgent desire to capture thoughts, feelings and imaginative pictures on paper. He had never felt that particular urge, and rather than sit in the shadow of fame he was very happy in his present occupation. After this brief insight he never again became personal, and spoke only about his present work as an instructor.

'The training courses at the College last one or two years,' he told us. The students are instructed in ordnance survey, map reading, detailed animal study and the use of firearms. Almost half their time is spent out in the field undertaking practical tasks. Actually, it can often be quite risky for the instructor.'

'What's the most dangerous thing you have to encounter, then?' asked Jill, pursuing her favourite theme.

'An African on a bloody bicycle,' someone shouted. Everybody roared.

'No, it's a green student who has never held a gun before. When he points the barrel straight at your middle and calmly pulls the trigger to show it's jammed, you tend to feel a bit warm.' There were more bellows of laughter.

'At the moment we're carrying out a reduction programme,' Pat informed us. 'The conservation work has been so successful, there are now over 20,000 elephants in Tsavo, but if nothing is done they'll soon be short of food. Of course, there's no need to waste the meat. Elephant cuts are far tastier than beef and I'm thinking of getting some of it canned for sale in the United States. After all, what's the difference between eating elephants or cattle?' He looked at his watch. 'Wow, we've just got thirty minutes, so let's move fast.'

While returning to the 'assizes' I realized how Pat's easy, warm personality had immediately dispelled my original nervousness.

There are few independently rich men who would be content with such a primitive existence, yet Pat Hemingway's life is devoted to preserving the animals of Africa. Perhaps his only concession to wealth is a personally owned twin-engined aircraft, which he pilots himself to log many hours on game counts and investigations.

Back in Taveta the judge eventually arrived at 3.30 p.m. Then, after Frank had been summoned to move his Toyoto which was parked against the assembly bell, a dull metallic clang rang out across the remote countryside and we packed into the small Court House. There were at least 100 others, and as there were not enough wooden benches many people had to squash together on the floor.

During the first hour, local cases were tried. The judge's English was translated into Swahili for the prosecution and into Kikuyu or Sukuma dialect for those giving evidence.

There were several cases of being drunk and disorderly. One man had urinated in a public place. A boy was accused of stealing a tattered jumper. Someone had made indecent advances to his ex-girl friend in a sisal plantation. An elderly father of a large family was accused of using poisoned arrows on his neighbour's livestock. How the translations fared in these cases can be imagined!

But at last the European was called. If found guilty he faced a fine of up to £10,000 and/or a maximum of five years in jail.

'This is a very serious offence for a good man,' said the prosecution.

The judge corrected the statement – 'You mean it is a very serious offence for a civilized man.'

There followed a detailed cross-examination. Unfortunately, the police contradicted themselves frequently. They also failed to call the Park Warden as a witness, and presented their evidence so badly that the judge had little option but to dismiss the case.

Some of the African rangers were extremely annoyed, and made wild accusations against the police force, although any deliberate lack of integrity could not be proved.

'Oh well, this all goes down to experience,' acknowledged Frank.

'It certainly does. Let's go back and drown our sorrows with a bottle of filtered water,' said Pat.

And we did just that.

CHAPTER VI

Field Studies in Tsavo

The wild life stronghold of Kenya – Aggressive lions – A reign of terror – Finding an oasis – Roaring Rocks – Tree destroyers – Complete annihilation – Status quo in the jungle – Water diviners – A form of sonar – Animal theories – Simple questions without an answer – No greater thrill – Ants and aches – Professionals and amateurs – Death with efficiency – Living symbols – Something repugnant – A voice in the wilderness – Guns or cameras – Heartening facts – Elephant noises.

FOR a few days after the court case we remained in Tsavo, the main wild life stronghold of Kenya: elephant, buffalo, rhino, hippo and lion, as well as many other mammals, birds and reptiles can be found here.

This National Park is well known for the fierceness of its lions, an aggressiveness inherited from their ancestors who were slaughtered extensively by slave traders following the Galana River to the hinterland with cargoes of chained prisoners, ivory and skins. Then the unexplored sanctuaries ruled by these 'kings of beasts' were penetrated by the construction engineers who came to build a railway from Mombasa to Lake Victoria.

As if sensing the threat to their existence the lions began a reign of carnivorous terror which for months stopped all progress on the railway; workers were even snatched out of the carriages to be devoured or mutilated. This long battle was vividly described by Colonel Patterson in his famous book *The Man Eaters of Tsavo*.

But they were eventually conquered by humankind's mastery of destructive skills. The railway was at last completed, and it brought prospectors, hunters and pioneers who exterminated animals in their thousands.

1. An elephant sifting thorn bushes along the Nile, Uganda.

Happily, Tsavo is now a conservation area, the largest in the world; and in spite of more than 1,000 miles of railtracks and roads, the scenery remains unspoiled, exhibiting much of the same savage glory which belonged to the darkest Africa discovered by Colonel Patterson over half a century ago.

Among the greatest attractions is Mzima Springs, an oasis of tropical vegetation surrounded by miles of parched land. Mzima has abundant water supplies, for millions of gallons flow unceasingly from springs which have their source far away in the northern Chyulu Hills. Like sponges these hills soak up the rain, which is then absorbed through underground channels of lava to emerge at Mzima as perfectly filtered crystal water. The deep pools of Mzima are fringed with tamarind trees. Monkeys *hide* amidst the overhanging branches and frighten observers by leaping upon them without warning, and hippos bask in the translucent water.

We discovered this lush oasis by accident. The vast plains reduce everything to a miniature, as if viewed through the wrong end of a telescope, and from a distance Mzima appeared to be no more than a few green plants growing in an enormous flowerpot. It was a striking experience when, at the bat of an eye, the arid bush changed into a cool land of fertility.

That same day we climbed the western slope of Roaring Rocks which, although only fifty feet high, looks out on its eastern side across a valley hundreds of feet below.

The scene was familiar. Scattered nimbus clouds floated in hyacinth blue. Hills and valleys were strewn with gneiss and schist rocks that had resisted erosion for 600 million years. Herds of buffalo, appearing no bigger than insects, caused puffs of red dust to spiral upwards on the hot air like smoke signals. A few vultures fell quickly from the sky in an almost vertical descent, legs extended, on to some carcass or dying animal.

Only the wind noises were different; a narrow gulley changed these from a sigh into a thundering blast. Indeed, Roaring Rocks was an apt name.

Pat Hemingway had told us there were more than 20,000 elephants in Tsavo, and when we settled down to photographing some of them it was easy to understand why a reduction programme had become necessary. Every year they require 800,000 tons of green fodder, and such food was obviously getting scarce. Feeding mainly on leaves, tender shoots, wild fruits, seed pods and bark, the elephants generally obtain these delicacies by ripping up the trees. They perform this feat by using their forehead to exert an ever-increasing pressure on the tree while at the same time employing a foreleg to spread the load evenly and so prevent the trunk from snapping before its roots are unearthed.

Frequently, we saw several elephants join forces to attack a really big tree and then some of them pushed as the others pulled. Once overturned, the foliage is quickly sifted with the efficiency of a phenomenal vacuum cleaner, any unpalatable pieces being spat out like grape seeds. Afterwards, the roots are levered from the earth and the bark is stripped as easily as removing tissue paper. Everything is eaten and only the framework of the tree remains; this lies broken and distorted like an aeroplane wreckage. Then, deprived of shade and protection, the soil quickly dries in the sun and is blown away like sand. There are now many square miles of Tsavo which have been trampled flat and left covered only with a litter of wood slivers.

However, away from the National Parks this plant annihilation can render a vital service in maintaining a balance in the growth of vegetation, thereby preventing many areas from becoming impenetrable jungle.

Another service is dowsing; in arid regions many animals owe their lives to wells dug by an elephant after locating an underground water supply. Some experts say that elephants achieve this divination simply by finding a dried-up river course and then choosing various sandy locations where they remain stationary until their colossal weight causes any subterraneous water to ooze out. Later, we saw this happening in Kidepo; but it hardly explained the numerous wells which are a long way from any possible river bed and surrounded by miles of parched land.

However, after our own studies, I came to the conclusion that elephants may detect water by using their giant ears as selective sound scoops and perhaps being able to differentiate wet or dry ground from the shock wave of their tread. This method would be a very credible form of sonar.

While exploring the dry Serengeti plains I watched a solitary bull elephant actually tapping on the ground with his foot and apparently listening; with endless patience he continued to repeat this operation in different places. Eventually, dusk forced me to leave but on returning early the next morning there were obvious signs of a hole which had been filled in. I scraped down in the sandy earth almost to elbow depth, and found it slightly damp!

Utilizing the same intuition and perseverance, these enormous beasts also dig for salt and laxative minerals to eliminate their many intestinal parasites.

Of course, there are many people far better qualified than myself to present theories regarding animal behaviour. Nevertheless it is surprising how little is known about this subject. What causes certain animals to migrate? Why do the baboon troops disappear for weeks on end? What is the longevity of lions, and over how many miles are they prepared to follow migrating herds of game? Is it true that giraffes are voiceless? To such questions there will one day be answers, and that is why so many people find it exciting to spy upon animal habits.

For me, there can be no greater thrill than to spy with a camera. To crouch on the ground with a blue African sky above. To breathe the acacia perfumed air. To be separated from danger by a mere bush. To remain silent and hidden. To wait. To observe. To keep the camera poised for instant action knowing there is generally but one opportunity to get a picture.

Looking through the viewfinder requires absolute concentration and all discomforts are forgotten – Hours of leg cramps, the vicious ants, the sharp thorns and the burning heat. Then *click* and it is over. Hours of preparation for the fractional part of a second.

Your body aches, your arms and legs are scratched, your clothes are torn and soaked with perspiration and the ant bites are itching madly.

Yet, although most of my illustrations in this book could have been made from the comfort of a Land Rover I prefer to believe that in some indefinable manner, photographs taken the hard way capture the sweat and excitement of the moment. Even if the prints are not perfect they still exhibit a certain atmosphere of authenticity which others often lack.

Even so, I have used a Land Rover on numerous occasions and I am certainly not decrying the animal pictures taken by tourists who enjoy the advantage of four wheels. These are amateur photographs, and while on holiday everything a tourist undertakes is naturally relative to the pleasure obtained. But a 'professional' is in a different category. Whether a photographer, writer, artist or musician, any professional captures the trust of his public and this confidence which is bestowed upon him entails a responsibility to produce the best work possible, irrespective of comfort.

All my book material is obtained by this guiding principle, although it so happens that nothing gives me such delight as striving to achieve complete authenticity and accuracy both in pictures and text. To hunt with a camera therefore offers me untold satisfaction. It is an exciting sport from which one never fails to learn a great deal about the characteristics of the animals being recorded.

By comparison, hunting with a firearm does not appear to be a sport at all, though such a 'sportsman' is defined in the dictionary as *one who undertakes a daring pursuit of wild game and will give his opponents a fair chance.* Yet safaris are now booked like a Cook's tour. The hunters set off in comfortable transport, accompanied by a team of staff and several resident professionals who know just where the game is to be found. Their equipment includes luxury tents (often fitted with air conditioning units worked from portable generators), furniture, comfortable beds, crockery, radios, record players and refrigerators. Food is prepared by expert chefs and because additional supplies can be ordered by radio-telephone any skimping is quite unnecessary. Naturally, the barman can mix a perfect cocktail. . . . This is how a sportsman undertakes the *daring pursuit.*

Out for the kill, every move is organized by the professionals: locating the lair, finding a safe hide, choosing the best firing position, hustling the game to ensure it moves in the required direction. Then the sportsman is ready for action, and he adjusts

II. Kilimanjaro from the Marangu Hotel, Tanzania.

the telescopic sights on his high-powered ·458 Winchester. Should the excitement throw him off aim, a professional takes over with an accurately placed shot for which he modestly gives his client the credit. And often, the victim is something like an impala, ostrich, hartebeest, topi or water-buck. . . . This is how a sportsman *gives his opponents a fair chance.*

It is natural for men to hunt but there is a fearful difference between the old methods used by natives and the modernized style employed by the white man. Steel traps, poisoned darts or wire loops may cause the animals greater suffering than a bullet, but such devices are so inefficient that hunting with them is a battle of true cunning, rather than a single cold-blooded action.

Rifles spit out death with a mechanical efficiency which tends to separate the marksman from nature, and leads him to regard life scornfully. Just as one nuclear device can destroy a city, so can one shot from a high-powered Winchester kill two buffaloes. The comparison is relative; where the profusion of wild life has always maintained a status quo against native traps and arrows, it has swiftly succumbed to the onslaught from guns.

Now, East Africa is about the only accessible wild life stronghold left in the world, and its wonderland of animals are living reminders of all that civilization has destroyed and can never replace.

I am not insinuating that *every* hunter who uses a firearm is skill-less or heartless. Over the years, many experts have matched their abilities against those of nature, and lost. But I find it impossible to recognize as sport – in the true sense of this word – any kind of hunting with a firearm. The luxury safaris are even worse, and to me they smell of make believe, of the pretender and the egotist.

As a competition marksman with both automatic pistols and rifles, I am not adverse to firearms. I can fully appreciate the excitement of cleverly tracking and shooting big game. But pursuing wild life for the mere sake of killing, and with practically all the odds in my favour, has always seemed repugnant. I wonder how many of us would attempt to penetrate the jungle on foot, and to eventually face a lion, while armed only with a quiver of arrows or a few spears. In these circumstances, would the so-called sportsmen still be at their clubs recounting how they daringly gained such magnificent trophies?

My own hunting is done with film and I have always advocated the camera instead of rifles. Six years ago I wrote the book *Underwater Photography*, pleading against the ever-growing use of spear guns, but it was a small voice which remained unheard in yet another wilderness. Now there are miles of European coastline where it is impossible for snorkellers to find even one colourful Mediterranean fish because their diminished numbers have been forced to hide in the safety of deeper water. They may never return to the inland shores.

Fortunately, the animal conservationists are trying to prevent this kind of disaster on land. Naturalist photographers also help by providing the material for wonderfully illustrated books such as *African Wildlife* – compiled by my good friend Franz Roedelberger – which have indirectly persuaded many people to exchange their firearm for a camera.

Particularly heartening was the fact that many of the most dedicated conservationists we met, begun their career as ardent professional hunters. These are traits which the trophy sportsman may like to ponder. Certainly no one could have a more renowned family background of big game hunting than Pat Hemingway.

The days around Tsavo passed quickly. When we left for Tanga it was in that dew-laden blackness before dawn, and the air was filled with the noise of elephants drinking at a nearby waterhole. I could hear the abrupt squeaky notes of those playfully spraying each other, and the vibrating rumble of enjoyment as they discharged a great deluge into their mouths with a swooshing gurgle. 'There's nothing like a trunkful of pure water to quench the thirst; five gallons of satisfaction and strength in every filling,' the advertisement would say! Staccato squeals of remonstrative anger came from a mother shouting at her calves, and loud trumpeting bellows indicated a serious difference of opinion between two bulls.

As the noises faded in the darkness I was sure that I also heard the spontaneous, infectious laugh of Pat Hemingway, which is now surely as much a part of Tsavo Park as the cry of its animals.

CHAPTER VII

Southwards

A tropical spectacular – Hatari – Washboard roads – Tricky bridges – Thorn trees – Vultures squabble – The king arrives – An unpleasant sight – Nature's dust-bin – Vindictive game – Old buffaloes – Lions meet an unpleasant end – Awe inspiring – Punctures galore – Ten men for one job – A dying city – Claustrophobia – Bats and beetles – Murderers – Religion or black magic – The Persian settlement – A strange omen – Mud or dust – Showering and scrubbing.

AFTER leaving Tsavo and passing through Taveta for the last time, it was still dark when we arrived at the border post and paid yet another car toll of 30s to re-enter Tanzania.

Dawn did not break for perhaps an hour. To our left, a range of mountain peaks magically appeared as luminous outlines. The stars faded, and second by second the sky lightened. First it became a solid roof of greyish tin foil, then a flat, white, mercurial surface, and finally a nascent covering of lustrous silver which split into a spectrogram of every conceivable colour and shade. Yet the land remained black and, as we were travelling towards the brilliant horizon, this tropical spectacular reduced the efficiency of our headlights to that of insipid candles.

Wallop! Bang! Bang! Momentarily, the unexpected impacts wrenched the steering wheel from my hands. While watching the sunrise, the good road had ended abruptly and we were now driving on a badly weathered surface. Although the deeper faults were indicated by small signs of *Pole Pole* (Swahili for slow) and Hatari (caution) none of these gave adequate warning because they were erected actually at the side of the holes.

Often I rounded a bend to find a great rift directly in front of us. Sometimes these obstructions were avoided by swerving, and occasionally I had enough time to reduce speed gently. But frequently I had to make a quick, violent application of the brakes which sent our luggage crashing around the car. While carrying three spare wheels there was little space to house anything in the boot.

We encountered our first 'washboard' section some miles further on. Rain, wind and sun had moulded the hard-baked sand into an iron finish of closely-spaced ripples. Driving over them produced such a noise and vibration we may just as well have been riding down steps in a milk float. When the ripples were more widely spaced I found that a speed of between 40 and 50 m.p.h. produced the smoothest passage; but immediately the spaces decreased, anything above 10 m.p.h. was quite impossible. Along these sections, even a crawling pace made the suspension units hammer with loud metallic thumps, and the Cortina would twist sideways, emitting dreadful clangs from the transmission and back axle. Under these conditions the traction disappeared almost completely and when climbing hills, I often had to 'slip' the clutch to prevent a stall. Considering the heat and the load this must have given the clutch plates a great deal of wear, but starting off again after a stall necessitated an even greater strain.

At many narrow bridges the need for caution was emphasized by wrecked cars which had left an ominous hole in the parapet. More than once we bounced dangerously near to scraping these walls, although at our low speed it would have meant only the embarrassment of scratched paint-work.

The ripples were always at their worst on either side of the bridges. Any attempt to steer accurately or hold the wheel tightly merely increased the terrible punishment being delivered to the car, and I found it far better to let the Cortina jump around with the minimum of correction. The few vehicles we encountered were also travelling slowly, so on most stretches the entire width of road could be utilized quite safely. I therefore hardly touched the wheel, often watching it spin as if controlled by an automatic device.

As we wanted to reach Tanga by midday, the washboard

conditions provided no time for viewing wild life though we stopped once when I spotted vultures dropping from height to the nearby bush.

These vultures are indirectly responsible for many arrests of poachers, for National Park Wardens can deduce from their type, number and behaviour the exact location and probable size of a carcass. They can also tell whether the killers are still present or somewhere in the vicinity. In fact it is rare to look skywards for very long without seeing a few vultures soaring, wheeling, quartering and gliding effortlessly through the air as they cover scores of miles.

Of course, what appears to the non-expert as a purposeful drop is often deceptive. These particular vultures may therefore have done no more than descend to water or to their nests, because even ordinary manoeuvres are easily misinterpreted when the birds concerned are flying high and scanning an enormous area of land below.

However, parking the car some way from where I believed the birds had landed, I left Jill behind and walked noiselessly along the sandy road before moving into the bush.

Being in a hurry I carelessly brushed against a wait-a-bit thorn and some of its formidable curved spikes caught my shorts. To prevent the material from shredding, each of the fish hook claws had to be removed carefully and this is how the tree acquires such an unusual name. (The wait-a-bit is a genus of the acacia tree and the Masai utilize its razor sharp foliage as barbed wire fences to protect their cattle from predatory animals).

While disentangling myself the silence was broken by a loud hissing, and once free I followed the source of this sound, to find six hooded vultures around the remains of a serval cat which lay at the foot of a skyscraper ant-hill. As the vultures tore at the cat's mangled carcass the colour of their heads, which obviously depended upon blood pressure, alternated between a dirty pink and deep crimson, and their large, ugly bodies were an unpleasant grey. The hissings were an audible threat as they squabbled between each other over the possession of a piece of flesh.

With my 35 mm Canon camera and a 135 mm telephoto lens

I was taking some good shots of these scavengers when a lappit-faced king vulture arrived. The wing span of this enormous bird must have been eight feet or more. With a pure white breast, long brown feathers and a grotesque face hanging in pink folds he made excellent close-up material.

The smaller, hooded vultures continued to fight over the carcass while the king sat apart as if eating carrion was beneath his dignity. But after watching disdainfully for several minutes it seemed to strike him that perhaps the carcass should be investigated. Stretching out his great wings, he made an ungainly hop before waddling intently towards the mangled remains. As he approached, the other vultures looked up nervously, and without any hissings they moved away from the body.

The king chose a dominant position on the spine, scrutinized the victuals with a critical eye, and finally showed approval by thrusting his beak into the bloody mass of intestines.

This macabre feast may not have been a pleasant sight but, without the vultures, stinking bodies would lie scattered around the countryside to pollute the air and become the breeding place for disease.

In many communities, vultures still remain the only effective substitute for modern garbage and sewage disposal. Nesting around native villages they eat up the most loathsome filth; some species even feed their young on regurgitated refuse. But these horrible feeding habits keep the places clean and hygienic; the scavenger birds are nature's dust bins.

Having been completely absorbed in taking photographs, my heart missed a beat when I turned around to find a lone buffalo grazing about thirty yards away. It is one thing to study buffaloes from a Land Rover or to watch them furtively from behind bushes. It is a very different thing to be standing close to one while unprotected and caught in the open. That is the moment when you realize how small and weak you are compared with the size and power of a 2,000 lb. animal. (An average bull carries horns which measure 30 to 40 inches along their outer curve. The record – from Kenya – is $56\frac{1}{4}$ inches with 44 inches between the horn tips.)

In any case, buffaloes are among Africa's most formidable

and cunningly vindictive game. Their only natural enemy is the lion and even the king of beasts can meet a highly unpleasant end unless several lions make a combined attack.

Nevertheless, like most beasts, the buffalo is generally peaceful and rather shy until molested or wounded, although a solitary old bull – such as this – can have a dangerously wicked temper. However, when old, their acute sight and hearing becomes impaired and now, being luckily down-wind, I walked softly away and he continued to graze without so much as a twitch of his broad, drooping ears.

It had not been worth my while to take any photographs because the once jet black hair had become greyish and his widely curved, down-spreading horns were badly worn and broken. While his sight and hearing may have been inefficient, a keen sense of smell always remains and it is doubtful if a single lion could tackle him successfully even in old age. For a photographic animal study he was a poor specimen, but as a splendid elderly gentleman he was still capable of inspiring awe.

Safely back in the car again, we continued on the washboard road which became progressively smoother, and finally turned to loose earth. The reddish-yellow dust swirled everywhere, changing our exposed limbs into earthenware shapes. Perspiration trickled down the forehead to leave clean rivulets like scars against the orangery-tinted skin. Crusty accumulations choked our nostrils and ears; our eyes felt gritty, and our voices rasped.

Yet animals enjoy the dust. They will take a dust bath at every opportunity, covering themselves with fantastic shades of yellow, orange, red, white or grey according to the local soil. In Tsavo Park all the elephants are red, while immediately across the Tanzanian border they are jet black. This chameleon quality confuses many visitors.

Dust, more dust, and even more dust is distributed by the wind to all points of the compass. It is an integral part of the landscape in which the very essence of Africa is captured. Although our own vehicle raised a continuous cloud of these suffocating particles I would not wish anything different. Here, automobiles and bitumen highways are out of place; but the dust belongs.

We covered over 100 miles before getting a puncture; but then within a very short distance I was forced to make three wheel changes. Not being eager to patch tubes by the wayside I deviated along a rough lane towards a small hillside community where I hoped to find a garage.

This turned out to be the village of Tangazo. No garage materialized, but on a piece of waste ground there was a petrol pump surrounded by chickens. It was not long before the pump attendant made an appearance, and helpfully he spoke a few words of English. In seconds the situation was under his control and immediately another ten willing workers were co-opted, even though the only tool at their command was a heavy mallet.

Fortunately, the standard UTC repair kit is beyond reproach, and I provided all the necessary equipment. One man took a wheel from the boot, two men each used a hefty lever to remove its tyre, one man extracted the tube, one man felt around the inner cover for anything sharp, two men fitted a new tube and utilized their feet to help get the tyre back on the rim, one man wielded the mallet, one man operated the foot pump and one man checked the pressure. Ten men for a simple job. It was enough to delight the toughest Union leader, except for the fact that the work was undertaken with efficiency and smiles.

Very quickly all three wheels were finished, leaving me with a twisted nail, some wire and numerous pointed volcanic flints which had torn the tubes badly. But despite such dreadful road surfaces our Firestone tyres were still in excellent overall condition.

Nobody tried to make capital out of this situation and the pump attendant told me, 'All for nothing. Help is fun.' Naturally I insisted upon a token payment, and this gesture of appreciation delighted them all. When we left, practically the entire village came to wave goodbye.

Tanga, the second largest town and port in Tanzania, is built on a lovely plateau which overlooks a bay set amidst coconut palms. But it is not developed as a tourist resort because it had always been the prosperous nerve centre of East Africa's sisal economy.

Unfortunately, the synthetic rope products of recent years have almost ruined this once thriving industry, and Tanga is now a dying city with many empty houses, obsolete buildings and acres of sisal fields lying waste. Ten years ago the attractive harbour, dominated by a memorial to the sisal pioneer Sir William Lead, handled a greater sisal tonnage than any other port; but now it lies comparatively idle.

The best hotel, with about twelve rooms, was falling rapidly into disuse, and offered few comforts or facilities. It catered largely for the overnight commercial traveller who sold agricultural products. The cooking was terrible, but at least the food came in quantity, and the other guests obviously had no complaints for they ate heartily with that enormous appetite typical of land workers.

After a hot, uncomfortable night we set off to visit the Amboni Caves. These vast limestone caverns, said to be an attraction unequalled in East Africa, lie five miles north of Tanga on the Mombasa road.

At first we appeared destined never to find Amboni, but in the hamlet of Kiomoni an old woman directed us along a bumpy track. It squeezed its way between a river on one side and towering rock formations on the other. Approaching the caves was like entering the yawning mouth of some gigantic beast. All the nearby trees appeared stunted by comparison.

Some people believe these caves stretch for 140 miles to Mombasa but there is little reliable information because most of the locals are afraid even to approach this area, and it is virtually unexplored.

''Allo, Bwana.' The unexpected voice came from a ledge above. It belonged to the resident guide who, I had been told, would give us every assistance. The guide informed me that his most important daily job, apart from providing lights was to rake the sand inside the cave's mouth. It was hard not to smile at the implied responsibility of this task.

'You see, only by keeping a smooth surface is it possible for me to detect the prints of any dangerous animals which often enter the caves,' he added.

My stupid smile faded, and instead I felt grateful for his serious disposition! As there were no paw marks in the sand we went through to the first gallery, which was about twenty feet high. From here the cave gradually narrowed to a passage where the low roof forced us to walk with bent heads.

The passage continued to shrink until it became no more than a rocky tube and like our guide we had to lie flat on the ground and crawl, with only a few inches of clearance.

As Jill suffers from claustrophobia she must have been finding this exploration particularly unpleasant. Even without any fear of confined spaces it was hardly an enjoyable experience, and I wished secretly that we had asked to see no more than the average tourist. Why *does* pot-holing have such a fascination for some people?

Thankfully the tube opened out so that we could stand again. Shining my torch upwards, I was surprised to see a very high, cathedral-style domed roof. Being totally dark the torch beam appeared as a veritable searchlight and its intensity disturbed hundreds of bats which flew around with the directionless speed of leaves caught in a whirlwind.

The air was full of these nocturnal creatures and the noise from their membraneous wings produced a musical buzz like a humming top. Turning the light downwards, I found the cave floor equally alive; it was covered with beetles, which were thriving upon a thick manure deposit of bat droppings. The guide assured us these insects were harmless, but it did not prevent us both from suffering a psychological itch!

Leaving the bat chamber we carried on to see other interesting sights. There was the hiding place of Osale Otango and Paul Hamis, two criminals who terrorized Tanga for years murdering anyone who ventured into the caves.

Another section contained magnificent stalactites, their icicle forms uniting with translucent pink stalagmites, both of which are lime carbonate deposits formed over the centuries by water seeping through limestone rocks.

The guide pointed out several places of worship where religious ceremonies are still performed though he was extremely vague about them. Fresh chicken bones, candles, used matches and empty bottles indicated that some kind of ceremony had recently taken place, but whether inspired by religion or witchcraft I could not say. We saw also a network of passages which have never been investigated because they are so steep and dangerous. Who can guess where these passages may lead?

Undoubtedly, the Amboni Caves hold many secrets and are a fascinating exhibition of nature's craftsmanship. Nevertheless, this particular fascination was not an essential part of the Africa which I had come to find.

There are caves all over the world. France claims the *deepest* explored cave where man has descended to almost 1,000 feet. The *longest* surveyed cave is in Switzerland and this stretches 37 miles. But the longest cave *passages* run horizontally for 150 miles and belong to the United States. We have seen caves in Thailand, New Zealand, Japan, India and South America as well as on the tropical islands of Trinidad, Jamaica and Tobago. So, the Amboni Caves excited no personal enthusiasm. Their dark attractions could never equal that wild glory which lives and thrives beneath the African sun.

When we finally emerged back into daylight, the flower-perfumed air, the vivid colours and the drowsy warmth were all sensory pleasures which we sampled as if deliberating the bouquet of a rare wine.

However, this disenchantment with the caves extended only to their physical characteristics. I am still enthralled by archaeology and while undertaking historical research in the West Indies I was thrilled to make exciting discoveries of tools, weapons, bones and pottery which belonged to the Arawak and Carib civilizations.

Because of my interest in antiquity we stayed in Tanga for another night, and visited Tongoni to see the remains of a Persian settlement. A popular legend recounts how the founders of this settlement came from Shiraz, after their ruler – Sultan Hasani bin Ali – had dreamed that his bed was being eaten by a rat with iron jaws. The Sultan interpreted this dream as a warning that his kingdom would crumble, and with many faithful servants he fled from Persia by sea. Eventually their boat reached the East African coast, and at Tongoni the Sultan founded his new colony during the year 1350.

Tongoni was a large community. To the northwest of the

site there is evidence of numerous stone houses, being part of an extensive settlement; and as the central mosque obviously had four pillars this building must also have been large. Around the mosque are forty intact graves and among them are the greatest number of pillar tombs ever found.

Although our day in Tongoni was worth while, I was pleased to be leaving the Tanga district. A dying town has a sad and empty atmosphere.

Passing the harbour for the last time, I looked northwaras towards an island which lies half a mile from the mainland. It is called Island of the Dead. A legend foretells death to any inhabitants and it has never been populated. Momentarily I wondered if this empty island was a strange omen of Tanga's collapsed prosperity.

On the Dar es Salaam road, apart from an alternating surface of washboard, loose chippings, dust and holes, there were also major alterations being done by an inexperienced East European contractor. Progress was over a year behind schedule, and the results so far achieved were reminiscent of a Laurel and Hardy project.

There was one specially memorable stretch where the workers had demolished fifteen miles of usable road, leaving in its place nothing but churned-up ground and a high centre ridge of soil, with enormous mounds of gravel, earth and bricks dumped at random. We snaked between them, following a crazy course more suited for an obstacle race. I was frequently forced to cross the central ridge, and as the car sheered off the top soil like a snowplough, in spite of knowing we had armour plate protection, I remained extremely tense, unable to reassure myself that a concealed rock would not mean a shattered sump. It took over an hour to negotiate this particular section.

By the time we arrived in Dar es Salaam we both had orange sweat rings around our eyes, ankles and wrists. There was a gritty lining between my toes, beneath my nails and even in my navel. Every piece of exposed flesh was covered with hard baked dust, and we emerged from the car looking like two effigies carved in rock. When a hotel porter went to empty the boot, an orange dust cloud enveloped him as if he had just raised the lid of an enormous powder compact. The luggage and the spare wheels were recognizable only by their shape.

'In Africa, guests arrive either covered in mud or dust,' the porter told us philosophically.

Surprisingly, no dust had penetrated our Revelation cases, and their surfaces were so far quite unmarked. If they survived this journey it would be a unique experience, for we had tried numerous other makes of case over the years, some much more expensive, and all of them were wrecked before our travels were completed.

However, Jill's watch fared rather badly, and on removing the rear casing I found the mechanism thickly coated in dust and some parts covered with a green fungus. But my own Rolex was still working sweetly. Only a first-class watch can withstand the dust, heat and humidity involved in our life of travel. I had already owned two watches of reasonable quality but, like Jill's, neither of them gave more than a year's service. The Rolex has now been ticking for several years in spite of dust storms, monsoons, and constant deep sea diving.

In our room we scrubbed ourselves clean with the same zeal which the animals and birds exhibit when they take a dust bath. I imagine that wild life's enjoyment at becoming covered in dust is inversely proportional to the satisfaction man gets by washing it away!

With the last orange particles swooshed down the drain we both gave a sigh of pleasure and relaxed.

CHAPTER VIII

Island of Political Intrigue

Two islands – Revolution – Armed guards – A Treaty – Malaria – Submarine invasion – The Chinese and East Germans – Closed – Making history – On a tightrope – Strange attitudes – The storm – Local specialities – Big Brother – Broken eggs – Slave traders – Antique doors – City by night – Banned literature – Cloves – An overshoot – A biplane – Striking and frightening – Tanzania's stability – A powerful weapon.

DAR ES SALAAM was founded in 1862 when the Sultan of Zanzibar decided to transform the small fishing village of Mzizima into a port and trading centre. This is now the capital of Tanzania and although generally referred to as *Dar*, its name in full means *The Haven of Peace*. Today, such a name is still justified. The city centre is built around one of the most picturesque harbours in East Africa, while attractively developed residential areas overlook palm-fringed bays.

As a city, Dar is typically African and much of the former atmosphere remains. The harbour is frequented by quaint Arabian dhows[1] which unload Persian carpets and other Eastern goods in exchange for spices and ivory. There is a considerably native shipping industry with old vessels plying between here and the numerous river creaks to collect mangrove bark, grain, copra and other tropical produce. At the busy market there are no tourist curios, though every type of African food is sold and the stalls attract not only the Masai but also a strange mixture of people who have Moslem, Christian, Hamitic, Nilotic, Bantu and Arab origin.

There are few animals to be found around Dar and the nearest National Park lies 180 miles away. Our main reason for

[1] A single-masted Arabian Sea ship of about 200 tons.

travelling so far south was to visit the nearby islands of Zanzibar and Mafia, both of which would make an interesting diversion before we moved to Manyara, Ngorongoro and The Serengeti, where wild life is at its greatest concentration.

Since Independence was granted to Zanzibar in 1963, a succession of revolutions and counter-revolutions have kept the island more or less closed. Even during periods of acclaimed calm visitors were often met by aggressive guards, who carried machine guns and conducted ridiculous interrogations.

But now, three weeks before our arrival, many restrictions had been lifted and for day trips, a visa – which normally took several months to obtain – was no longer required. East African Airways were making daily flights to the island and it was claimed that a special Treaty had once again united the two States as the *United Republic of Tanzania and Zanzibar.*

On paper everything sounded fine, yet government officials in Dar were uncommunicative about the situation. Who contrived the most recent revolution? When was the Treaty negotiated? Which Party was controlling Zanzibar? Nobody appeared to know the answers. The Tanzanian National Tourist Board were equally uninformed and mysteriously declined to provide us with any literature because it was forbidden to distribute leaflets which were printed before the revolution. Who had given this order and why? The question was unanswered.

In spite of this Treaty with Zanzibar the scheduled flight there was classified as an international route, and demanded a departure tax of 15s each. We left at 7.30 a.m. on a DC–3 and the forty-five miles were covered in less than half an hour.

At Fumba airport there was a lengthy form to complete and an entry charge of another 15s each, but everyone was friendly and I could see no armed guards. Customs and Immigration were operated courteously and so was the Zanzibar Friendship Tourist Bureau, which presented everyone with two Chloroquine anti-malarial tablets and a glass of water. Apparently, Zanzibar is relatively malaria free but the disease is very prevalent in Dar, and this gift was an effort to control the number of mosquitoes becoming reinfected. However, as we took regular prophylactic doses of Paludrine I refused the offer.

Except for requiring a pass to visit the beaches (we may have been directing a submarine invasion force!) there was no restriction on our movements, and we found a tranquil island with blue lagoons and white sand bays.

Once in the capital[1] we began to discover that the Chinese and East Germans had taken control. There were Chinese-built factories utilizing Chinese-trained employees. Schools had been erected and staffed by East Germans. The curator of the Peace Museum had replaced certain historical exhibits with an extensive model showing Chinese methods of agriculture. Shops were selling goods produced mainly in communist countries, and even the two kinds of beer available were bottled in Peking and the German Democratic Republic. Kilimanjaro, the beer of Tanzania, was unobtainable.

The Waterfront Palace belonging originally to His Royal Highness the Sultan, the House of Wonders (built by the Sultan in 1883), the Arab Fort dating from 1784, the Residency and the Local Handicrafts Showrooms had all been closed to the public and were guarded by soldiers. Additionally, the entire population now have to carry identity cards and they are not free to leave without authorization.

Dust had gathered on the empty shelves of the Tourist Information Board. The Indian manager explained that the Revolutionary Party had destroyed all literature because the history of Zanzibar could no longer be recognized, and a new version, more acceptable to the Party, was being written for the benefit of visitors.

By holding Zanzibar, the communist powers have a definite foothold in East Africa, but it is difficult to see how this affects Tanzania. President Nyerere accepts a great deal of communist aid but as Tanzania is a poor country likewise he takes help from the West with equal alacrity. Currently, the projects under Chinese and East German direction are more or less balanced by those instigated with American and West German resources.

But China's influence is unmistakable Its staff of around

[1] *Author's Note:* Although not recognized generally, only the capital is called *Zanzibar* and the island's true name, as used by the Zanzibaris, is *Unguja.* But to avoid any confusion, when I speak of Zanzibar, I refer to the island as a whole.

5. Beautifully marked zebras in Ngorongoro, Tanzania.

6. A giraffe leaps across our track, Ngurdoto, Tanzania.

The author in a Masai village, Tanzania.

eighty high-ranking officials at the Embassy in Dar must be a record. At one time these officials were free to wander throughout Tanzania, but now President Nyerere has restricted the movement of *any* Embassy staff and before diplomats can get permission to leave Dar, the reason for their proposed journey must be approved.

Although it would appear that President Nyerere is walking a very tricky tightrope between communism and capitalism, the general opinion is that he will keep his balance.

While we spoke with some Zanzibaris who appeared enthusiastic about the new regime, others, after looking furtively around, drew us aside and became almost hysterical in their bitter condemnation.

This is perhaps to be expected, but the most difficult attitude to understand was that expressed by European residents who maintained that 'everything in the garden is just fine', and laughed at the very idea of Zanzibar being under communist domination. Yet, when asked any pertinent questions, their typical reply was: 'There are things here which we don't talk about. Zanzibar has received a great deal of exaggerated publicity and now we must wait for our tourist trade to return'.

It will take a long time to rebuild the travel industry of the island. Travel agents quickly remove trouble spots from their list, to protect clients, and to prevent the personal irritation of frequent en bloc cancellations. But many people retain their faith in the island's future, and at the Zanzibar Hotel numerous renovations had just been completed.

This is a small, unpretentious hotel of great charm, old but spotlessly clean. The accommodation is simple but comfortable, with facilities such as air conditioning units and hot water geysers. The private bathrooms are antiquated but practical; although the bath water empties directly on to the floor and sluices down an open drain, the system works efficiently.

We had planned to arrive at the Zanzibar Hotel for a late lunch, but soon after 11.00 a.m. the daylight vanished with the suddenness of a photoflood lamp burning out. The sky became the smooth colour of black velvet and a squall blew up, turning the sea into a vat of dark foaming Guiness. It rained heavily.

This was the dry season, and the storm should have been over quickly, but after waiting under cover for an hour it had merely intensified to a blinding deluge and the road was flooded to a considerable depth.

Still optimistic about the weather, we decided to make an early lunch, and with difficulty a taxi got us through the narrow streets which were like canals.

The carved, brass-studded Arab doors of the hotel were just three fast leaps away from the taxi, but in those few vital seconds of exposure the cataract did its worst. The British manager and his wife could not have been kinder, and as well as offering to lend us dry clothes they placed a room at our disposal.

For lunch the chef prepared *Kiku Wakupaka*, a delicious local speciality of chicken cooked in deep coconut milk, and served with rice. Afterwards we ate *Andazi*, a popular type of bun made from ghee, flour and sugar, as well as the inevitable coconut milk. More than satisfied we moved to the reception lounge, where Jill read a book and I watched the downpour.

We were not alone; two Chinese men sat talking in hoarse whispers. Instinctively their secretive manner attracted my curiosity and I glanced across at them just as they had decided to scrutinize *me*. Our eyes met. I nodded and gave a half smile. They continued to stare without any acknowledgement. Turning my attention to the rain I gazed outside for several minutes, but then temptation was too strong and once again I looked quickly in their direction. They were still staring. It was a suspicious, unfriendly stare. I contemplated my nails and felt decidedly unnerved by this *Big Brother is Watching You* act. I never risked another indiscretion, and was relieved when shortly afterwards both men left.

With their departure only one person remained, a young local who endeavoured to appear a Cuban Bolshevik type by sporting a beard and a black leather jacket. Sitting at a table piled high with empty beer bottles, he was decidedly drunk. When eventually he strode away with an unsteady gait, he was oblivious to the fact that eggs were dropping from a brown paper bag held carelessly beneath his arm.

I shouted a warning; the receptionist shouted; the porter shouted: but our warnings failed to penetrate the alcoholic haze and, leaving a trail of shattered yolks, he carried on. Dropping a final splurge at the main entrance, he stumbled down the steps, regained his balance, gave a friendly wave and disappeared into the rain.

The man who came to clear up the mess told me that since the revolution all laws imposed by the previous Arab rule have been reversed. Anything inspired by the Mohammedan faith is now taboo and consequently, with no time restriction on the sale of alcohol, some people are drinking too much. Fortunately, the novelty is apparently beginning to wear off and even our egg scatterer – who was a regular customer – gets drunk far less frequently.

The storm continued to rage, and at four o'clock a telephone call from the airport informed us our flight was cancelled. Fortunately there were plenty of vacant rooms at the hotel and as our entire luggage comprised three cameras, we remained in the lounge where the manager joined us; as he had once been a pilot, we passed an hour or more exchanging stories about flying light aeroplanes.

By 7.00 p.m. the rain had eased and the stars were visible. Water still poured from the roofs and swirled along the roads as we paddled our way round the town.

Zanzibar was once among the biggest slave trading centres in the world. Its main thoroughfares are still no wider than alleyways, and without footpaths there was little room for a car to pass us safely.

The barred lower windows of the small whitewashed houses are a reminder of those days, less than a century ago, when the slave dealers and pirates paraded the streets.

Although the houses are of little architectural merit, many of them have a magnificently carved, brass-studded door like the one we had first seen at the Zanzibar Hotel. These are genuine Arab-style antique doors set in a square frame with every stud, fashioned elaborately from solid brass, sparkling with the same brilliance it exhibited during the times of slavery.

A few doors, possibly dating back to the very beginning of Zanzibar, have additional brass spikes which, according to historical notes, were intended as a deterrent to war elephants and battering rams. After this protection became unnecessary the lethal points were reshaped to a harmless boss end, and consequently none of the functional heavy spikes remain in existence.

Apart from being a remarkable decorative feature, these antique doors embody Zanzibar with tremendous character. Tragically, such character belongs to an era which the Revolutionary Party will not recognize, and the most ornate doors in Zanzibar – decorated with gilded texts from the Koran – are in the *House of Wonders* where entry is forbidden.

Now, as we walked beneath the stars through wet, twisting streets, the town had come to life. Shops had thrown open their shutters, and tailors sewed by the light of paraffin lamps. Women sat making baskets while children played. The coffee vendor, carrying a great metal pot attached to a bowl full of glowing charcoal, attracted customers by clanging two metal cups in a catchy melody. We were told that his tune had been the same for thirty years, and it was certainly more enticing than Toni Bell or Cadbury's 'hot chocolate' advertising jingle. The coffee was delightful, with a rich bean fragrance typical of the *Arabica* variety; served piping hot it fortified us against the damp air.

Often a road would narrow to no more than a passage and then we became tightly packed in a motley throng of Indians, Africans, Somalis and Persians who all showed interest in our presence and gave friendly smiles.

Prior to the revolution, bearded Arabs were conspicuous in the noisy crowded streets, with their long, dark blue robes trimmed with gold braid, coloured turban, and the customary sword or dagger at their waist. But even without these spectacular figures the town had the romantic quality of the *Arabian Nights*. However intangible this atmosphere, it is one which the Revolutionaries will find hard to destroy.

The cancelled flight was a stroke of luck for it had given us the opportunity to experience the fascination of Zanzibar by night. As we had managed to buy essential items like toothbrushes, there was little inconvenience from our unexpected overnight stay.

The next morning was sunny. During breakfast the airport

telephoned to confirm seat reservations on a flight leaving at 4.00 p.m., which gave almost a complete day at our disposal; we made full use of the time to visit many historical places, including the tall, white house from where Livingstone planned his last expedition into central Africa.

Christ Church cathedral, lying in the crowded bazaar area, was erected on the site of the world's last open slave market. Its altar stands on the site of the whipping post. The crucifix on the chancel pillar is carved from the tree which marks Livingstone's grave at Chitambo. Although the Revolutionary Party are discouraging visitors to Christ Church, we were ushered in by a verger who even gave us a banned historical information pamphlet; it seemed quite innocuous, but was classed as defamatory to Zanzibar.

Later we toured the countryside where cloves are grown extensively, every plantation acre having 60 to 70 trees, each with an annual yield of approximately 350 lb. of dried buds. This is the chief export, and although the average British household may regard cloves as annoying woody pieces found in apple pies, they are also used for the manufacture of antiseptic, confectionery, flavourings, toothpaste, perfume, liqueurs, soap, tinctures and even Javanese cigarette tobacco. As the locals were often chewing the green clove buds we picked several, and found they stimulated the palate with a tingling freshness.

Most of the plantations were completely waterlogged. According to the newspaper, the previous day's storm had been the worst for two years, with 3 inches of rain falling in the first four hours. More heavy rain was falling as we arrived at the airport and half the runway was obscured by the low misty clouds.

Shortly after our arrival I heard the DC-3 on its landing approach. The variation of engine power indicated the difficult conditions and when a grey, shapeless mass with two shining eyes eventually materialized from the gloom, only an overshoot was possible. Then both wing lights were extinguished, the gear began to retract and she climbed away steeply. The second approach was no better, and after this the engine noise faded away as the plane returned to Dar.

The meteorological office expected no improvement for

several days, so we jumped at the offer of a lift back in an old-fashioned Dominie biplane, which could always get through when the larger aeroplanes were grounded. Taking off was relatively simple, and according to radio reports from Dar the landing conditions were fine.

The storm seemed to be concentrated within a small area, for after we had climbed bumpily to 5,000 feet the sky became suddenly clear as if someone had pulled back a roller blind. Zanzibar lay hidden beneath a lake of black ink, but we landed at Dar in sunshine.

Our trip had been mentally illuminating. To see a tropical island taken over completely by a communist-inspired revolutionary party is a striking experience. To realize this situation is not the fictional plot of a novel is frightening.

Unbelievably, Zanzibar has been reopened to visitors. From my own discreet enquiries it seems that the Party want the island to remain inaccessible, but as President Nyerere regards Zanzibar as a powerful tourist attraction, he exerted pressure on the Revolutionaries. His apparently successful negotiations say much in the favour of Tanzania's continued stability.

Similar conditions to those in Zanzibar have sealed off numerous islands and countries behind an impenetrable iron curtain, so it may be a good thing for as many people as possible to visit Zanzibar.

While entry to the island remains possible, hope for its future stays alive and any tourist development can be a powerful weapon.

CHAPTER IX

A Date with Barracudas

Orderly development – Big fish – Incredible stories – A tricky landing – Trolling – Barracuda – Silver missiles – Capture evaded – Fighting hard – A large rock cod – The trial of strength – Foul hooked – A strange argument – Shades of heroism – Sentiment or obtuseness – At the controls – TWA entertain – Threats – Book critics – Television with Spike Milligan – Sour grapes – Snake charmers and tribal dances – A 500-mile journey.

MAFIA, lying eighty miles south of Dar, has no connection with the hostile Sicilian mafia and, unlike Zanzibar, it has gained a reputation for orderly development. Situated almost at the mouth of the Rufiji River, the island's eastern reef is about 750 yards from the Continental Shelf where the floor of the Indian Ocean plunges down to 100 fathoms. Big fish come from this deep water to feed upon prawns washed down by the Rufiji, and there is an abundance of barracuda, king fish, red snapper, wahoo, tunny, rock cod, sail fish, marlin, shark, five-fingered jack, and species of the caranx.

Some of the biggest tunny known in East Africa have been caught at Mafia, and several world records for other varieties were won here, including one gained by Sir Arthur Conan Doyle who, in the Songa Songa area, took his 75½ lb. record with an 80 lb. line.

According to the stories we heard in Dar, there is nothing extraordinary in getting either a 100 lb. caranx or a 1,000 lb. sail fish, but accounts of incredible fishing grounds circulate with the same speed and doubtful accuracy as a racing tip from the stables.

I know of many famed spots around the West Indies where it is quite possible to cast or troll for an entire week without any strikes. Experts had once assured us that on an uninhabited island in the Pacific the fish practically came ashore to say 'hello', but while sampling a few days of complete isolation there, I may just as well have exchanged my rod for a baton. Much to Jill's disgust, we were forced to eat tinned food and coconuts until the boat returned.

As the Fishing Club at Mafia has only a grass air strip we flew out in a four-seater, single-engined Cessnar flown by Graham Stein, a young Swiss. Although an experienced pilot, this was his first time in Mafia, so before undertaking a positive approach he made a dummy run, passing over the field at fifty feet with full air speed.

Even for our Cessnar the strip was exceptionally short, and it appeared as a few yards of narrow green carpet lying between two palm-covered hillocks. Obviously Graham was used to this kind of landing and immediately he swept round on the final approach. Expertly controlling the descent he dropped us over the palms, levelled out, momentarily opened the throttle to clear the hillock and then touched down, having completed the tricky approach without any variation in air speed.

The grass carpet disguised a hellishly rough surface and this spoiled our copy-book landing with a series of unnerving jolts.

The Fishing Club is next to the air strip, though Fisherman's Inn would be a more appropriate title. It is small and quaint, untidily attractive with a fishy charm. Pinned to the walls are faded photographs of proud fishermen showing their catch, and reference books on fish are scattered over shelves and chairs. There is a tank of tropical fish and the skeleton of a barracuda. A turtle shell is propped against the wall, and piles of fishing magazines are stacked in a corner.

But nothing is contrived and this stimulating, casual atmosphere evolves from the character of Arnold Klosser, an ardent fisherman who is the manager. He lives alone at the Club and welcomes the infrequent guests like personal friends.

Only minutes after our arrival we were aboard the *Kiwi*, a 29-foot cabin cruiser with a 35 h.p. diesel engine. We trolled three lines.

'Round the point, over the rock shallows there'll be barracuda,' Arnold told us. 'If you have no objection I'll pull in a few to give the local boys, and then across the bay, as soon as the tide turns, we'll get some caranx and king fish for our own lunch. Use the red lures for barracuda, the brown spotted ones for king fish and the yellow for caranx.

He made everything sound so very certain, and I hoped the fish knew the exact rules. What if some hungry barracuda should gobble up a *yellow* lure. Or worse, a barracuda looking for the rock shallows. What a calamity that would be – a barracuda caught in the *bay* and on a yellow lure. And even worse, it may be the one Thursday during the year when barracuda fancy 'greens' and even the red lure would fail to tickle their palate. Or getting really basic, perhaps *all* the fish were having a 'green' day and nothing would gobble up anything.

This musing ceased abruptly as soon as we rounded the point. Z-z-z. Z-z-z. Both the port and starboard lines were running and we landed two barracuda totalling 36 lb. Out went the lines again and in came more barracuda; naturally they never touched the yellow lure on our centre pole. The operation ran like clockwork.

'Tide's on the turn. Let's get across the bay,' said Arnold. He dropped his great bulk into a fighting chair and began exchanging the two red lures for yellow. At that moment it was easy to visualize him pitting his obvious strength against an equally powerful fish, for those thick, mahogany-tanned, hairy arms were surely as strong as they looked.

Zz-z-z. Br-r-r. Zz-z-z. This time all three lines were running at an incredible rate. I dropped into a fighting chair and began the battle. I had always believed that only game fish such as tarpon jumped, but this was a tasty caranx species (what else!) and it shot from the water like a silver missile.

Already I wished that my arms were twice as thick. Sweat poured into my eyes, and once again the caranx broke the surface in an arc of flashing silver. For a second I failed to keep the tension and suddenly the line hung limp. My fish, which must have weighed at least 50 lb., had evaded capture.

The next caranx was not so lucky. Almost 65 lb., it gave a

tremendous battle before it was finally hauled in. Arnold told me this particular species, weight for weight, fights as hard as anything in the sea. I believed him. So did Jill, who without help landed one weighing 30 lb., although in the process her face turned the colour of a tomato and she perspired and puffed as if about to explode.

Just before turning back, we saw Arnold in action. Zz-z-z. The noise sounded identical with previous strikes but Arnold jumped excitedly.

'That's a big rock cod and it's going deep. I'll take this.'

How could anyone recognize a fish from the zip of a line? Yet if Arnold said this was a rock cod, then it would be a rock cod. I had no more doubts concerning the abilities of this amazing man.

'You've got to bring these in fast otherwise you lose them under the rocks,' shouted Arnold.

Now, yard by yard, we saw a trial of strength. He crouched forward, dropping the pole low and quickly winding in the slack line. Then, using his body as a lever, he straightened, at the same time raising the pole and throwing more weight against the cod by leaning backwards. Arm tendons rippling from the strain, body pouring with sweat, he nevertheless repeated the action non-stop with the smoothness and speed of a metronome beating rumba time.

Suddenly the cod broke free and the fight was over. 'I thought it would get away,' Arnold said. 'It was almost certainly foul-hooked and when that happens even the small fish can play hard. This one must have been 100 lb. or more.' He flexed his wrists.

Then and there I decided I would never make a real fisherman. To pull a small, spiritless 3 lb. mackerel out of the sea is one thing. To fight with a gallant, shimmering silver caranx is something else. Why? The question is impossible to answer. Logically, if a person tries to catch fish it should make no difference whether he catches a mackerel or a caranx. Both are good to eat.

Perhaps I am too sentimental. After all, if someone hooked *me* on a line, I would jump and battle to get free. Survival is a natural instinct not a matter of heroism. But what if the 'someone' who hooked me was a being from outer space with superior intelligence and weapons. Unable to fight such an aggressor on equal terms, many spectators would surely find a touch of heroism in my natural battle for survival?

Anyway, it was impossible for me to regard these fighting fish merely as *Tritons among the minnows*, and whether sentiment or obtuseness, I found myself endowing their combat with a mark of bravery.

Even so, in one and a half hours we caught 174 lb. of caranx and notwithstanding the sentiment I ate my lunch with relish. We left shortly afterwards for it was already late afternoon and Graham wanted to fly back in daylight. He let me take the controls for our return journey to Dar, and this made a pleasant end to the excursion.

On our last evening in the capital we were invited to an Inaugural Flight Gala given by Trans-World Airlines. It was attended by several hundred travel agents who, at the expense of TWA, were enjoying a flying holiday with sightseeing tours arranged at Dar, Nairobi, Kampala and London. These particular agents were extremely rude. Above the rowdiness it was impossible to hear the welcome speeches and guests were constantly arriving late and very drunk. Afterwards, dancing to the hot tempo of African drums, the gala turned quickly into an evening of noisy revelry.

I cannot imagine why airlines persist in entertaining travel agents. It does nothing to increase business, and market research has shown that when the public enter an agency, their choice of holiday resort has nearly always been decided in advance. Besides, few agents know anything about the world. Just listen to their hedging if you ask for a specific recommendation to meet certain holiday requirements, or enquire if they can plan a special tour, or simply expect them to make reservations for a personally devised schedule.

But to enter their office and say, 'I want to stay for ten days in Davos', is a slightly different matter; you are then sold a *twelve*-day package tour. Should the original request have been for ten days in Montana, then you still finish up with a twelve-day package holiday in Davos because 'hotel rooms are difficult to get in Montana, the communications are bad, it would mean an overnight stay somewhere en route and in any case the latest train schedule is not available'.

It is generally the travel writer, not the agent, who excites the desire to visit a particular place. Books and articles honestly written, provide a service not found among commercial interests. Unfortunately many travel writers are not independent, and the few honest Johns often get hung for their pains.

My book *Eastern Quest* came under heavy fire from the Government of India Tourist Board, who threatened an injunction to prevent further distribution unless it was re-written. Eventually this futile case was 'dismissed', but the book has been banned for sale in India, while according to verbal sources I am now classed as an 'undesirable alien' and will never be allowed to return there. In Jamaica the local newspaper ran a full page feature about *Tropical Quest*, saying 'Get out your cutlasses for Townsend'. One of my recent European Guides was boycotted from the 'book list' of a certain British hotel organization, even though officials admitted my disagreeable facts were completely accurate! The Tourist Board of Fiji is exceptionally displeased with me, and so are similar Boards in half a dozen other places, although fortunately *most* countries would welcome our return.

But where the travel industry may play rough, most of the book reviewers have been extremely fair and I am grateful to them for their criticism and praise. During the worrying legal situation over *Eastern Quest* I am sure the critics helped considerably, for within three weeks the book had received 103 major reviews, all of which gave credit for a horrifying but obviously unbiased and truthful report of a journey through modern India. On the day of publication the London *Evening News* ran 22 column inches entitled 'No Medals for the New India', and the book gained coverage in papers from the *Financial Times* to *Titbits*. Even ATV producer Bob Gillman asked us to talk about our Indian experiences, and together with that great personality Spike Milligan, who was born in India, we were interviewed by Gwyn Richards.

Naturally, the travel industry as a whole gave *Eastern Quest* a cool reception, for they prefer the authors who compile cooing-phrased, cliche-filled guides.

TWA's arrangements for food and entertainment were well planned, particularly the buffet which included a wide range of Tanzanian delicacies. There was pumpkin with coconut, sweet potato and fish, corn-cob in coconut oil, curried fish, fish with beans, boiled maize and porridge with casava leaves. (These dishes remained almost untouched by the travel agents, while hamburgers disappeared by the plateful!)

After dinner the Wamakonde people gave an impressive exhibition of tribal rituals. This was followed by Wasukuma snake charmers who danced while draped with some of Africa's deadliest snakes. These Wasukuma came from a special Mwanza fraternity which claims to have a powerful medicine giving complete immunity to snake bites. At the end of their performance, two of the cobras were milked of venom by squeezing of the poison sacs.

Then, not to be outdone, a number of bourbon-soaked travel agents took the floor, grabbed the amazed Wamakonde dancers for partners, and to a rapid drum-beat commenced a let-your-hair-down fertility riot. By 11.00 p.m. the floor was packed with uninhibited revellers, and it was no hardship to leave early.

The next day we were driving to Monduli, an insignificant hamlet where the Regional Administration for Masailand is based. We intended to visit the Area Secretary to obtain his official blessing on our plans to visit the remote Masai villages. Should we encounter any trouble, this courtesy call would stand us in good stead.

Hoping to reach Monduli by nightfall – a journey of almost 500 miles – we were setting off before dawn.

CHAPTER X

Mud to Monduli

Skating – Disasters – A near miss – No margin for error – Stuck – Murder – The silent warriors approach – Disconcerting news – Waves of vibration – Fuel runs short – Road signs forty years old – The God of Fire is benevolent – Cyclists enthused with power – A threat and a crash – Spontaneous laughter – Our mobile mud house – Human intimacy – African glory – Meeting the Area Secretary – The final warning.

THE storm in Zanzibar had not been the localized climatic freak I had believed. On the way to Monduli, miles of earth road were nothing but oozy mud, providing skids and thrills galore.

Holding the steering wheel very gently I skated rather than drove, occasionally slipping sideways. We saw a heavy lorry which had slithered into a ditch and now lay at an oblique angle with mud up to the axles. Further on, there were two cars and a bus in the same ditch.

Suddenly an army truck was approaching from the opposite direction, travelling much too fast. Sensing danger, I pulled close to the bank. My fear was justified for the driver suddenly lost control. The truck spun around completely, skidded towards us head-on, then slewed around to pass within inches while facing the wrong way! I had a quick glimpse of a laughing black face with teeth as white as an elephant's tusk. Most Africans have no conception of peril until the inevitable actually happens, and the laugh was quite genuine. But the nervous reaction from a moment such as this always leaves my own legs feeling like cotton wool for several minutes.

Where we negotiated precipitous sections I held my breath, realizing that any slight error of judgement or failure to correct a 'drift' could easily send us over the side.

Along one dangerously hilly section the car became stuck, but luckily a number of Masai were gathered on higher ground watching for occasional cars with the same morbid interest shown by spectators who vist the *Wall of Death* at a fairground.

These Masai were all of 6 feet 3 inches tall; standing erect, with their spears held upright, they cut an impressive silhouette against the skyline. Only when it appeared certain that our Cortina was absolutely bogged did the warriors come to help.

Naturally, when they began moving towards us en masse, I was unable to guess their intentions and felt uneasy. After all, there are good and bad people in every community; these could easily be the bad! I knew that a month ago, in this same area, a tourist driver had been speared to death for taking photographs, and those responsible were so far undiscovered.

But although this crime was drastic, I also remembered that the passengers in the car admitted the driver had ignored frantic protests at the use of his camera. Some exceptionally primitive tribes have come to believe that cameras are weapons which extract blood from their bodies, and officials say the murder was therefore a tragically misguided effort at self protection.

Still the Masai came nearer to us. Their faces were expressionless. Their movements silent. Their walk deliberate. Their manner proud. Their spears and shields producing an occasional glint of hard steel.

Assuming what I imagined to be a casual attitude, I stuffed our camera equipment beneath some luggage and then got into the driver's seat and began slightly revving the engine as if trying to move again.

'I think you're scared,' Jill said.

'Scared about getting stuck in a bit of mud?' was my contemptuous answer.

'No, scared of *them.*' She nodded in the direction of the Masai.

'Oh, *them.*' I tried to sound as if I had just seen them. 'How lucky to have people around. They'll be coming to help us.'

'Is the gun loaded?' she asked ominously.

'It's always loaded, but what do we want a gun for?'

'I was just thinking about that driver who was speared,' came the disconcerting reply.

'That was his own fault. Besides, I doubt if the Masai had anything to do with it. None of the natives like having their photographs taken.'

'Yes, but you can never be sure how primitive tribes are going to react. Their values are different from ours. And we're miles and miles from anywhere.'

Of course, Jill was right, even though I pretended not to agree with her. The primitive tribes do have different values, and I thought about yesterday's newspaper item which had been headed *The Indignant Cannibal.* This was reported as follows:

' "I killed and ate only members of my own family, so who was I offending?" a villager protested to the magistrate at Njombe District Court in Southern Tanzania.

The court had listened with rapt attention to the macabre tale told by a 50-year-old Mahola Petikite, who pleaded guilty to harbouring instruments of witchcraft and admitted to killing and eating the flesh of his father, wife and three children. He said he was unaware of any criminal aspect of his actions.

Petikite said that some time ago he had not been well and approached a witch doctor for medical treatment. He had been given several plants, some of which he planted near a pond and the others he mixed with fowl's blood and claws.

Through this medicine he had developed a craving for human flesh. By using witchcraft he killed and buried his father, later exhuming and eating the body. Petikite continued: "I killed and ate my wife and children in the same manner but with my family gone and no more human flesh available – I was not empowered to eat anyone else – my physical condition failed."

He went to another witch doctor who advised him to destroy the plant near the water if he wanted to be rid of his craving. Instead, Petikite said he simply transplanted it and when the witch doctor had found that his advice had been disobeyed he accused him before the Matamba Primary Court of harbouring witchcraft instruments.

The case was later transferred to Njombe and now the file

7. On the way to Meru, Tanzania.

The light was failing when this hyena snatched a piece of zebra, Ngorongoro, Tanzania.

8. The author, Pat Hemingway and Hans Reinwalf outside the District Court, Taveta, Kenya.

An antelope showing its donkey-like ears. Along the Nile, Uganda.

has been passed to the High Court for a decision whether Petikite should only be sentenced on the witchcraft charge or further charged with mass murder.'

But different values or not, any uncertainty about the behaviour of these Masai was needless, for after thrusting their spears into the ground, without speaking a word, they pushed us free. Once again we were squirming and sliding our way through the mire.

Where the mud finished, a washboard surface began and the corrugated iron effect had been made very much worse by the storm. Shock waves passed through the chassis, turning the car into some kind of infernal vibrator apparatus which threatened to disintegrate. The ripples pulled in all directions, causing the steering wheel to shake madly and transmit unpleasant oscillations. I gripped it tightly, fearing that the whole car would break up at any moment.

My next worry was for petrol. Since covering the first 200 miles without finding a single petrol station, three subsequent pumps had been dry. Now we had emptied our last reserve can, and although I switched off the engine to coast down every hill it was not long before the gauge showed empty. Then an Esso sign appeared. Then a BP sign. Then an Agip sign. With salvation apparently near at hand I paid no more attention to fuel economy and accelerated, confident that petrol would be available at the next village.

I should have known better. We had already discovered that African road signs are unreliable. Special warnings such as *Floods*, *Cliff Fall* and *Obstruction* become a permanent fixture until they are removed fortuitously by the wind or a peevish elephant. But even though the rainy season is heralded with gales and the elephants are invariably peevish, a battered notice-board outside Moshi had still warned of 'Railway Extension Work' – the railway referred to was the Mwanza Line which opened in 1928!

The petrol hoardings proved to be equally unreliable and there was no Esso, BP or Agip filling station. Instead there was a single Shell pump and beside it a man sat treadling a sewing machine. When I asked for petrol he stopped treadling and laughed, blurting out a tale of woe which sent him into further paroxysms of mirth.

Apparently, this pump was a new installation. To economize, the management had utilized a second-hand tank which now leaked and must be repaired. The repair was not a simple job because the tank lay buried deeply under a new concrete fore-court. Also, the pump worked from electricity and due to lightning the village power supply was out of action for another three months. In any case, petrol deliveries to this location had always been spasmodic, so the technical problems were really of little consequence. However, if I wanted a suit made, the order would present no difficulties. He began to treadle again.

Luckily, countryfolk are ever helpful. They may tell you something is five miles away when it is nearer fifty. They may top up the car battery with swamp water. They may lose the radiator cap. They may commit a dozen outrageous indiscretions, yet if sanity can be preserved recompense is always forthcoming, for their hearts are in the right place.

The tailor was no exception, and after I showed him the fuel gauge with its needle registering zero, he promised to return with petrol in five minutes. We waited. Ten minutes passed. Thirty minutes. Sixty minutes. After one and a half hours the prodigal arrived back somewhat drunk and puffing a tattered cigar. Nevertheless he was clutching a couple of two-gallon cans filled to the brim and splashing petrol everywhere.

The cigar agitated me considerably, especially when he blew through it, and removed the ash in a shower of sparks. I was afraid that at any moment a disastrous flame ball would overtake our benefactor, but obviously Vulcan felt benevolent. Indeed, *all* the gods must show equal charity towards the Africans, otherwise their lives would consist only of misfortune, for these people tempt fate wantonly.

With slurred voice, he apologized for being a 'few minutes' late. Then, standing the vital cans precariously on the Cortina's roof, he extracted from his pocket two bottles of Coca-Cola, and prized off their metal tops on the front bumper. Sh-shwoof. Sh-shwoof. Twin jets of warm, sticky, fizzing Coke shot over us as the erupting bottles were thrust into our hands. Graciously we sucked at the remaining froth.

But for the petrol I was truly thankful and having expressed my appreciation in words and money, we drove off, hoping to make up lost time on the better road. Chance was not on our side and soon another delay occurred with what Pat Hemingway's crowd had described as the most dangerous thing in East Africa: 'an African on a bloody bicycle'.

Just as Londoners indicate their success by owning a television and a car, the badge of success in a small African village is a sewing machine and a bicycle. The sewing machines produce a fair amount of needle-butchered fingers but at least only the operators suffer, whereas natives on a bicycle are a danger to anyone in the vicinity.

Bush roads see little traffic, and a solitary cyclist, perhaps riding for the first time, becomes enthused with power and an ill-conceived confidence. Imagining the road belongs to him alone, he wobbles on, swerving from side to side, lost in the sheer thrill of performing these manoeuvres. Faster and faster; the speed is hypnotic and exciting. Suddenly, the reverie is broken by a car horn which the cyclist interprets as a challenge to his supremacy. He pedals harder and swerves more violently. Then, feeling the engine breathing its heat down his neck, the illusion of absolute power is finally shattered and either he falls off or makes an unexpected, suicidal turn to escape. I doubt if any of these cyclists realize how badly they scare a motorist!

More experienced riders can be just as dangerous when they carry outlandish loads: bundles of bamboo which stick out for several feet on either side and occupy most of the road's width; sacks full of fruit balanced on the cyclist's head; bales of grass; furniture; metal roof sheeting. Everything imaginable is transported on these humble two-wheeled vehicles. Alas, the Africans believe firmly that if a bicycle is made to be ridden, then ride it they must, no matter how cumbersome or heavy the load being carried.

However, our skirmish took place with an unladen novice who was experiencing that first hypnotic thrill of speed and power. When we first saw him he was just a wavering blob in the heat-shimmering distance. The blob gave no cause for alarm and it could easily have been a vulture or a stray buffalo. Then, in a trice, the threat materialized.

Pedalling madly, a violently swerving cyclist was speeding towards us with his head down. On a straight road sloping gently for several miles the bicycle had gained considerable impetus, and there was no opportunity for any planned evasive action.

I slammed on the brakes and kept my finger on the horn. As the cyclist looked up his mouth dropped open in horror; he applied both brakes, and miraculously shuddered almost to a halt before the bicycle retaliated by shooting its rider into the air like a stone from a catapult.

Crashing onto our bonnet the cyclist stared at us through the windscreen with an expression of utter fright and bewilderment. Then the daze cleared and gave way to a wide, unrestrained grin. I helped him off the bonnet and he started to laugh. Nobody was hurt and the bicycle appeared undamaged. Once again the gods had been kind and when we left, the man was still laughing.

This whole-hearted laughter of the African is as spontaneous and inexhaustible as the Mzima Springs. The people are always prepared to laugh at themselves; and the absurd situations and ridiculous accidents which happen from day to day provide a never-ending source for merriment. Those who are incapable of facetiousness should not visit East Africa, for here it is essential to view life through the eyes of a humorist.

The last ninety odd miles to Monduli were on a fast tarmac road. In spite of our earlier buffeting, the Cortina had suffered no mechanical damage, and on this smooth surface she handled perfectly without so much as a rattle, although exterior-wise we resembled a mud igloo. Truthfully, we felt proud of our little car and affectionately christened her Goliath, which as well as meaning 'giant', also happens to be the name of an enormous, white striped African beetle that has tremendous stamina and courage.

Considering our long wait for fuel, the bicycle incident and the exceptionally tough driving conditions we were lucky to be so close to our destination by evening.

At this late hour we were both weary to the point of exhaustion. Neither of us wanted a night in the car, and as accommodation seemed unlikely to exist at Monduli, we seized the opportunity to stay at a native rest house. This was situated a few miles from Monduli and identified by a chalk written sign which bore the pretentious but mis-spelt title of 'Hottel'.

Before retiring we drank a bottle of warm beer and ate a frugal meal by the light of hurricane lamps. Our bedroom was a partially screened area containing odd pieces of heavy, coarse furniture and not even the convenience of a dressing table or a mirror. The carpet was rough concrete; cold water taps and a lavatory hut were outside. Yet, lying in my sleeping bag and gazing up at the loosely thatched roof, I felt an overwhelming contentment.

Through the bamboo slats came the high pitched whistle of tree frogs, the intermittent throb of crickets and the buzz of insects. Far away I heard a hyena laugh. The ineffectual screening offered no privacy, and from near at hand the loving murmurs of human intimacy blended with the jungle noises.

Low on the north-western horizon, Orion, a brilliant constellation of stars, clearly depicted the hunter with a belt and sword, while Orionid, stretching from the meteor system, gave the sword a radiant point. Just where the black forest met the clear sky, Sirius, known as the Dog Star or Orion's Hound, blazed magnificently through the unshuttered window opening. Two eagles flew past, their shapes outlined in silver grey. On the cool, flower-perfumed breeze I detected the faint roll of drums. This was truly African glory and well pleased I fell asleep.

The next morning we were in Monduli with the Area Secretary. At first our expedition plan was viewed unenthusiastically; but after a long talk the Secretary decided we were genuine people and not likely to make trouble. He then explained vital customs to which strangers must adhere when entering a Masai community, cautioned us about taking photographs, and specifically warned me against sleeping with the Masai girls. (I supposed it was in order for Jill to gallivant with the men!)

Finally, we were given a map of Masailand and a rough outline of how to reach certain villages. From here on, we would be alone.

CHAPTER XI

A Proud Tribe in Danger

Migrants – An intermingling of blood – A unique race – Relentless pressure – Fire-scarred landscape – The strongest people known – A vicious circle – Breaking with tradition – New facts wanted – Civilizations disappear – Mechanical failure – Getting lost is simple – On foot – An amusing story – Thorn trees – Gifts – Approaching the village – Proud dignity – Courage – Lion-killing rituals – Flies – Cow-dung houses – Two lusty nipples – Ideal fumigators – Sex customs and virility – Porridge – Colour superstitions – Tribal misfits – The price of happiness.

THE Masai tribe originate from inhabitants of the Middle East who, since the beginning of time, have migrated southwards. Moving through the African continent they caused all those before them either to retreat even further south or to be overrun and integrated with their culture. Huge forces of these migrants settled along the upper reaches of the Nile; but after remaining contented for numberless years, there sprang from this community a society of cattlemen who once again pushed south and east in search of better pastureland.

It was inevitable that, during their wanderings, these gentle Nilotic people cohabited with the fierce Hamitic tribes who populated Ethiopia, and interbreeding resulted in the birth of a new race. The men were tall, slim and arrogant, combining the tranquil pastoral nature of cattle-loving folk with the savage, barbaric inheritancy from warrior tribes.

Their women often had great beauty; and although possessing the courage to fight off an attacking buffalo, they were soft, affectionate, particularly desirous for sexual attentions and very proficient in the skills of love-making.

This intermingling of different bloods had therefore created a semi-nomadic Hamitic warrior people with their own temperament and way of life. This vital, hybrid race soon recognized their superiority in battle, and the immense power gained from owning vast heads of cattle. It was not surprising that they swept through East Africa, vanquishing the Bantu tribes and seizing all their domestic livestock while at the same time assuming control of an enormous territory which boasted the finest grazing lands. Here, then, was a race of fanatic cattle lovers and great warriors – the unique Masai.

Today, the Masai are still proud and brave, yet slowly and relentlessly they are being squeezed into extinction by civilization. Political and economic issues have taken from them much of the land they conquered, and their wanderings are restricted. Even when drought strikes or the grass is suffocated by thicket, they are not free to search out the pastures which lie beyond these enforced borders.

Too many Masai cattle are therefore chasing too little grass and this robs the soil of minerals, after which it quickly becomes infertile and the area eventually turns into wilderness. Waterholes are overcrowded and cattle even die from thirst.

Following a custom of centuries the grass is deliberately burned to encourage a fresh, nutritious growth, but whereas years ago this new growth was allowed to reach a sweet maturity, hungry cows now snuffle up the shoots almost before they have sprouted from the ground. Consequently, the process of regeneration is halted, no more rich grass is seeded by nature and a black, fire-scarred landscape combines with the scrub to form further unpasturable areas.

Crops have no value to the Masai, and their staple food remains milk, cattle meat and blood tapped from the jugular vein of the cow, and drunk while still warm. This unusual diet actually made them among the strongest people known, but as the grasslands decline, the health of their cattle also deteriorates and similarly the Masai themselves are gradually becoming weakened.

It is a vicious circle, and unless something is done the day is surely fast approaching when the Masai will fade from the African scene. What does it matter whether they survive or not? I doubt if there are many of us who care an iota one way or the other. But this *should* matter.

The Masai are victims of industrial and commercial expansion which is sweeping across continents and changing the whole pattern of life by destruction. Travelling constantly I see this destruction increasing everywhere and in just five years many beautiful countries have been altered so drastically that little remains of their former character. And as changes take place, as nature is mercilessly repelled by all kinds of commercial enterprise, so the undeveloped civilizations are likewise strangled.

But at present, having been the pulse of East Africa for centuries, the Masai are undoubtedly as much a part of the local scene as the lions, the acacia trees, Lake Victoria and Kilimanjaro. Although among the most primitive of races these people have a great intellect, and if they were now educated to live in harmony with *modern* Africa this would indeed be an estimable accomplishment.

Since the advent of widespread schooling, some Masai have broken with tradition to become doctors, lawyers, chemists, but they are exceptional cases. Large numbers of children never attend school, and those who are sent generally achieve little and develop no lasting scholastic interest. Nevertheless, the Masai possibly know more about cattle than any race in the world, and their horsemanship is unrivalled.

(Later, while we were in Keekorok, the horse belonging to an important local civil servant fell sick and after the veterinary surgeon had proclaimed the illness as incurable, an ochre-painted warrior was fetched from a nearby village; he treated the horse for several days and it fully recovered.)

Unfortunately, normal academic teaching can do nothing to help the Masai, but a special education programme is required to show them how fresh prosperity lies ahead only if they take an active part in wild life preservation, in conserving the rich soil, in helping the grasslands to flourish, and in cultivation. Owning useless herds of livestock is no longer a criterion of power and where twenty million cattle once meant salvation they now mean disaster.

If an intensive campaign was organized for Masai indoctrination and the same facts taught to their children at school, then eventually new leaders may emerge, the zest for life could return and a worthy civilization would have been saved.

Of course, the problem has enormous implications that are impossible to solve overnight but any attempt to find a solution appears unlikely. During talks with both Tanzanian and Kenyan government officials, I found an attitude of complete indifference to this matter and the Masai were usually regarded as 'a scourge who will finally disappear into oblivion'.

However, as East Africa has recognized the need to preserve wild life, I hope sincerely that the governments concerned may also revise their opinion about the Masai, whose only failing – if it can be termed a failing – has been a tenacious stubbornness in clinging to a former way of life. But throughout history many great people have been known for having a tenacious streak and a stubborn spirit. Such traits often go hand in hand with greatness, and I believe that if the Masai were helped sensibly they would contribute much to Africa.

To see a complete resurrection of this tribe would give me great pleasure for these primitive natives gained our affection and, after leaving Monduli, they welcomed us with respect, hospitality and friendship on numerous occasions. Distressingly, the 'ifs' which govern their survival are too big, and thinking realistically but sadly, I believe the Masai will be left to 'disappear into oblivion' like the Australian Aborigines, the Polynesian, the New Zealand Maori, the Melanesian and a dozen other civilizations. Is this really twentieth-century progress?

The Area Secretary of Masailand had been extremely helpful and we decided to make our first real acquaintance with the Masai by looking for a particular community some miles west of Monduli. We followed the Manyara road and as instructed found the 'turn-off' without difficulty.

From here we drove along a track which soon became obstructed by rocks, holes, anthills and other mounds. While negotiating a deep earth fissure there was a dull scraping noise from underneath the Cortina and we came to a sudden stop. The engine continued to rev and the gear was still engaged but we could not move. On releasing the accelerator we stalled. After restarting the engine, when I let in the clutch, Goliath shook all over yet remained absolutely stationary.

Fleetingly, I imagined every kind of complicated mechanical

failure, but having jumped out, one glance solved the mystery. Our sump was resting squarely on a mound and we were jammed with all four wheels off the ground.

Both of us set about rocking Goliath free but she refused to be launched and I realized how a skipper must feel when he grounds his boat. What an absurd break-down! Eventually, by using the jack I partly dislodged the sump from the mound and then a violent start pulled us clear. Only the Cortina's armour plating prevented a real catastrophe.

After this mishap, the obstructions were less hazardous, and instead of rocks, long springy grass, over six feet tall, topped the car. Ploughing through like a harvester we emerged ultimately on to a wide plain where an escarpment terminated our mobile travel.

Across the escarpment was open savannah and as Masai communities are perfect examples of unintentional camouflage, the binoculars were needed to verify positvely our discernment of a village; through the field glasses this turned out to be natural landscape. Absently, I continued to scan the countryside and by fluke I located the village in a slightly elevated position approximately two miles away. To the naked eye it appeared more like vegetation than the vegetation!

Before setting off to walk, Jill made a rough sketch of the terrain and I took an accurate compass bearing. Two miles may not be very far, but getting hopelessly lost is easy where every ant-hill or acacia tree looks the same as a hundred other ant-hills and acacia trees. At least we were unlikely to stumble across a lion because animals keep well away from these villages.

My greatest concern was about leaving the car in such a remote position. Even on a main road, an abandoned vehicle is often discovered by primitive natives who slash off the tyres and utilize the rubber as shoes. Even worse, the 'discoverers' may be educated natives who know the value of accessories and engine components.

Indeed, the Tanzanian Highway Code strongly advises people to avoid deserting a car in the bush, and while we were staying at the Marangu Hotel a European had recounted an amusing experience which endorsed the advice.

A day earlier, this man was driving from Mombasa when the

rear tyre went flat. Having raised the bonnet – to get at the tool kit – and jacked up the car in preparation for changing the wheel, he decided to sit beneath a tree and have a hot beer with some dry sandwiches. The shade was cool and he fell asleep for over an hour. Refreshed, he then ambled back to the car and started removing the wheel. Before the last nut had been unscrewed a noise from the front of the car attracted his attention, and there was a Masai warrior disconnecting the battery. 'What the hell are you doing?' he asked, hoping his Swahili was understandable. The Masai replied in perfect English spoken with an Oxford accent. 'If you're taking the wheels then I'm having the battery.'

Nevertheless, abandon the Cortina we must. But parked in long grass and still covered with the filth of our journey from Dar, she was well camouflaged as a mud hut or a termite's nest. Looking back, I doubted if anyone would recognize such a mound for a car.

Thorn trees were the things which made our trek to the village difficult. Thorns. Thorns. Every bush and tree has thorns. It always amazed me to watch the elephants vacuuming up this type of foliage; they must have iron palates. Admittedly the thorns are spat out, but sucking an orange full of tin-tacks would be preferable to sifting these wicked spikes. Hook thorns, talon thorns, short and long thorns; Africa is the continent of thorns and for causing discomfort they share an equal place of honour with the dust.

It was not surprising that we felt some trepidation at the prospect of making our first intimate contact with the Masai. Naturally, we hoped for a friendly reception, and to indicate our good intentions we had brought small bags of sugar, maize flour and sweets (for the children) all of which are prized commodities making valued but inexpensive gifts.

There were about thirty houses enclosed by a circular fence of thornbush in which a narrow entrance had been cut. At this entrance two men were sitting in the shade of an acacia tree and flicking away the flies with switches made from black wildebeest tails. This would be the chief with his headsman, and respectfully we stopped several yards away until beckoned closer.

Neither of them spoke English, but my previous months of

battling through a Teach Yourself Swahili suddenly became worth while. There are over 100 different languages in Masai-land and the chief spoke a bastard form of Swahili but at least the gist of what I said was understood.

Both men stood and extended their arms, resting a hand on each of our shoulders in welcome. The chief had a gaunt face wrinkled like a vulture's, and his arms were shrivelled laths. Although an old, old man, his body was held erect with the proud dignity of a leader.

The headsman was much younger, with horribly mutilated eyes. I gathered this terrible disfigurement occurred when he saved a cow by fighting off a lion with his bare hands. Today's warriors are equally courageous, and, in spite of conservation laws, if the king of beasts should wander near a village a youth will be given the opportunity to encounter the lion at close quarters while armed with a single spear.

Years ago lion-killing was an important Masai ritual to signify puberty[1] and, although this custom is now strictly forbidden, their art of skilful hunting has not deteriorated. All lions fear the Masai instinctively; in Ngorongoro we were amazed to watch a lone Masai stride through the grass towards a cowering lion which then turned and ran!

As the chief escorted us into the village we were immediately descended upon by swarms of flies attracted by the cattle dung that is always spread on the ground to discourage ticks. Consequently flies invade every Masai community and it is impossible for strangers not to become nearly crazed by their nuisance.

While the elderly folk often use a wildebeest switch, younger people generally make no attempt to brush off the flies. They allow them to cluster over their mouths, bunch around their eyes, creep up their noses and in their ears to seek out microscopic food particles or skin excretions of salty moisture.

All the babies in this village were partly hidden beneath an aggregation of flies, and small children ran around with their

[1] Days before the ritual, trackers would go out to find a simba (lion) and during this time the young initiates drank magic brews which were supposed to give extra courage and strength. Eventually, the lion was surrounded by about twenty naked boys and those proving their manhood had first either to jump astride the King of Beasts or pull its tail.

unclothed bodies covered by a seething mass. Even the women's broad pink nipples turned black and where a small girl sat naked I watched the flies actually crawling in and out of her vulva.

The chief came to our rescue and provided us with two fly switches. By waving them violently we prevented the buzzing cloud from settling.

At first the people were shy and regarded us with suspicion but after lengthy formal introductions the atmosphere became less strained. Groups of people gradually slipped away, only to return adorned in their most colourful beaded collars and arm bracelets. This was a nice gesture and typical of the Masai.

I obtained consent to take photographs and was surprised to see the teen-age females making sure their breasts were fully covered. A particularly beautiful girl caused me great amusement. With a perfectly formed body, sophisticated breasts and long shapely legs, she was as naturally luscious as a sun-ripe mango and as wild as the red-headed love bird which wings through the jungle tree tops.

When I raised the camera this girl pulled a thin robe over her breasts, clutching the material tightly and defiantly as if I might be tempted to snatch the garment away. But the cloth was half-rotten, and being strained across the fullness of such a voluptuous figure it tore just enough for two lusty nipples to emerge triumphantly.

The tiny boys were also concerned about their nudity, and when Jill offered some sweets to a toddler who could have been no more than five years old he dropped them in the dung while endeavouring to hide his little mauve penis.

Masai homes are made from cow dung and mud. When dry, their colour is a nondescript grey, and with curved roofs and no window spaces they resemble Anderson air-raid shelters. During the day we were invited to enter one of these houses.

Apart from the glowing embers of a fire it was pitch black, and without ventilation the heat and smoke took our breath away. Although decidedly uncomfortable, the house was an ideal fumigator and not conducive for insect life. The cow dung repels ticks and jiggers[1] while the smoke kills off flies; only the

[1] A tropical flea which burrows into the skin. The uncorrupted name is *chigoe*.

Masai are willing to 'stick it out'. It was hard to decide whether I preferred the smoky heat or the flies.

The house was divided into four compartments. A sleeping area for the wives, separate quarters for the children and calves, an exclusive section for the husband, and the cooking/dining room where we were sitting around the fire on three-legged wooden stools.

Straw covered the floor of each sleeping compartment and the husband had the additional luxury of an animal skin. Every night he chooses a wife to make love with, but once intercourse has taken place it is customary for the woman to leave him and sleep next to the children. If the husband is still virile he will then fetch another of his wives, and one warrior told me proudly that when dawn came the children's quarters were always overcrowded! (And to know when the dawn breaks, there is a hole – no larger than a penny – cut into the bedroom wall.)

A large pot of bubbling porridge was cooking over the fire and we were given a bowlful. Unsweetened, and mixed with flies which had succumbed to the smoke, it hardly stimulated our appetites but we ate every scrap because real porridge is an expensive treat for the Masai and their generosity could not be spurned.

When we returned to the fresh air and sunshine our eyes were smarting and watering from the smoke, but the sudden daylight made them worse; we squinted in actual pain as if facing an unshielded arc lamp.

When the time came to bid everyone goodbye the chief courteously sent a boy to show us the easiest way back. The car was still there and quite safe.

This first meeting with the Masai was one of many, and between Monduli and Ngong we visited scores of similar communities. The villages were identical except for trivial differences such as a greater or lesser concentration of flies and a varying degree of prosperity.

Everywhere the Masai were primitive but hospitable. Whereas most African tribes are affected by at least a veneer of civilization, the Masai have never allowed their lives to fall into any incongruous pattern. They retain the individuality of the same daily routines and tribal customs they have followed for centuries.

Primitive tribes that endeavour to become civilized take a long time to fit into another mode of life. A typical example of this was recorded by an educational officer who decided to distribute portable radios among a few educated local tribes.

The first tribe refused the receivers. 'What! Blue cases? You must be crazy. Surely you realize that if we have blue our wives will become frigid. Get us red instead.' The educational officer had no answer to this, so he hid the blue radios and offered to get the next tribe some red models, but at the mention of red they recoiled in fright.

'If we have red, our cattle get swine fever', he was informed seriously.

He changed his tactics for the third tribe he visited, and invited them to choose a suitable colour. Apparently blue or red was admirable, but almost any other colour would cause a giant water snake to rise from the nearby lake and eat up their children.

At least the Masai are not tribal misfits, and they never attempt to disguise their primitiveness. Radios, bicycles, white shirts, ties, sewing machines, cigarettes, beer parlours and all the other native badges of progress are generally disregarded. The Masai tribe still belongs to wild Africa.

Perhaps this accounts for the people remaining happy. The Masai have largely ignored the march of progress, and we found their villages marked by peaceful coexistence, strong family ties and an integrity of purpose. For this alone, civilization threatens the Masai with the supreme penalty.

CHAPTER XII

Lake Manyara

Tussle with a lorry – In the Great Rift Valley – The African bush telegraph system at work – Like a shower of jewels – Imperata grass warns of danger – Our first dik-dik – Tsetse bites – Sleeping sickness – An unproved theory – Resistance to poison – Infection checks – Modern drug cures – A leopard strikes – Hyenas and lions smell blood – Days of leisure – To Ngorongoro.

AFTER several nights at the rest house just outside Monduli, we moved on to Lake Manyara encountering an incredibly harsh road. Jagged stones and large rock chippings were spread deeply over fine yellow earth. Burst tyres were scattered around liberally, and every few hundred yards a glistening patch showed a windscreen had shattered.

According to the map this dreadful surface ran for about seventy miles and from the outset we were plagued by the dust of a vehicle ahead of us. For a time I avoided the dust stream by staying far behind but this reduced our speed to a crawling pace. I guessed the vehicle was a heavy lorry.

We stopped, ate a sandwich, took some photographs and then set off again at a very moderate speed. But we soon caught up the lorry because of the long hills.

The only solution was to overtake, a hazardous and difficult manoeuvre.

Resolutely I drove into the lorry's dust cloud and began to close the distance between us. Visibility decreased to almost zero and was just as bad as any London pea-souper. Soon I was driving completely blind at 55 m.p.h., frantically straining my eyes for the dark shape which I knew must materialize in front of us at any moment. There it was! I braked instantly and hard, readjusting our speed to keep the shape in sight.

III. A Masai woman with her child, Tanzania.

It was impossible to see if the road was clear and I followed the overtaking procedure recommended by experts: 'Stay close, spread your fingers on the screen to prevent it splintering, say a prayer and wait for a gust of wind to blow away the dust.' I stuck close and Jill held both hands on the screen.

Stones were flying towards us like machine-gun tracers and some of them actually hit the screen. Expecting the glass to break at any moment I was on the point of giving up the tussle when a blustering cross-wind whirled the dust over the countryside and momentarily I saw a mile or so of clear road. Unhesitatingly I pulled out and accelerated through the murk into sunshine.

Lake Manyara National Park spreads out from the foot of precipitious cliffs in the Great Rift Valley, and though a compact area of 123 square miles, it contains a great variety of scenery and wild life. The park's main all-weather track begins one and a half miles from the village of Mto-wa-Mbu.

When we came to the ranger post and gave our name there was an unusual scurry of activity.

Having meticulously recorded the Cortina's number, the warden then asked me to print our surname on the same piece of paper. We were then told to wait until the man returned. I racked my brains, trying to think what crime we may have committed. Perhaps the cyclist who crashed onto our bonnet had caused trouble? But that was days ago.

My reflections came to an end when the warden arrived back with another man. This man was a Masai, and it transpired that his brother lived in one of the villages we had visited around the Monduli area. He merely wanted to shake our hands.

How had this man heard about us? How were we expected at Mto-wa-Mbu? We never learned the answers. Monduli was over seventy miles away and all the Masai villages had been isolated; mules had been the only form of transport available. Besides, it was an impromptu decision to stop at Manyara; our original intention had been to make direct for Ngorongoro.

The incident left me in no doubt as to the effectiveness of Africa's bush telegraph system, and I remembered what Dr von Nagy had told me about hidden eyes.

More than two-thirds of Lake Manyara is within the National Park and its shores of reed beds, marshland and mud flats provide the ideal breeding ground for birds. Along the water's edge we were overwhelmed by the confused jumble of fluttering vividness, like a firework barrage: purple, yellow, orange, brown, carmine, white; speckled and mottled; irradiant and iridescent. Others sparkled in amethyst, ruby, emerald, sapphire and garnet, flashing through the air like a shower of jewels. There were so many species here.

At a single glance we could see herons, jacanas, waders, egrets, ibises, pelicans, storks, wigeon ducks, Egyptian geese, hammerheads, gulls, kingfishers and dozens more too numerous for naming.

Flamingoes floated on the lake, forming a wide pink raft at least a mile in length. Wagtails added a splash of yellow to this pink scene by hopping from one floating plant to the next, and thereby walking over the water as easily as land.

The early explorers of East Africa wrote an extravaganza of praise for the flamingoes: 'A ribbon of flame'; 'water set alight with fluttering wings'; 'wings that formed curving beaches of soft pink sand', are just three of the many recorded descriptions. The world has approximately five million flamingoes, and more than half of them live in Tanzania and Kenya.

The hours passed quickly and after exposing several rolls of film the sun was getting low and we returned to the car. Driving back towards the proper track we got lost in tall, broad bladed imperata grass which had attractive, white, fluffy seed-heads. This type of grass usually grows in muddy soil and often indicates a shallow drainage channel that spells trouble for the cross-country motorist, so I felt quite anxious until we rejoined the track.

Back in the forest areas and the open glades there were elephants, waterbucks, baboons, gazelles, warthogs and zebras but none of the famous Manyara lions or leopards. But we saw our first dik-dik, which gazed at us inquisitively. This was a female and as if wearing very high heeled shoes she crossed the track in a series of dainty little hops, coyly displaying her slender legs.

Almost at the same time we made another first encounter, for tsetse flies invaded the car. Their bites were most unpleasant and as well as drawing blood they bruised the flesh like hypodermic injections given by a lethargic matron at the outpatients department of a typical National Health Service hospital.

As a carrier of *nagana* or sleeping sickness,[1] the sinister, greyish-yellow tsetse is a malignant pest throughout tropical Africa. There are many infected areas where the flies kill off all the cattle as well as exposing humans to infection. During 1901–6 the disease ravaged Uganda and over 200,000 people died.

While Africa's wild life has become immune to the whip-shaped blood parasites (trypanasomes) causing this ailment, certain experts think the animals are a reservoir of infection. They say that as the tsetse feeds on the blood of an immune antelope or similar animal, the trypanasomes undergo a cycle of regeneration, and then the tsetse will be transferring deadly parasites to any non-immune livestock it bites. However, the destruction of game as a means of controlling nagana is controversial and no real success has been achieved during closely supervised experiments. Every year huge sums of money are spent on research into tsetse control yet nothing has succeeded so far. The flies will simply vacate one territory to invade another.

Initially, the powerful modern insecticides promised complete eradication, but this tough pest quickly developed a resistance to such poisons as DDT, DDVP, pyrethrum, pyperonyl butoxide and similar compounds. Now, the older methods of selective bush clearance – leaving only the trees and undergrowth known from experience not to harbour the flies – are being integrated with chemical treatment and spraying from aircraft.

Even the *Harris Tsetse Catcher* devised thirty-eight years ago in Zululand has been revived. This is a sacking-covered wire frame trap. But although these traps have captured two million tsetses there are certain species of the fly which are never fooled.

Fortunately, the tsetse fly has a low rate of reproduction. It

[1] Nagana is the cattle disease corresponding to sleeping sickness in man. *Sleeping* sickness must not be confused with *sleepy* sickness.

lays no eggs, but hatches a single egg within the abdomen and extrudes a living larva that changes into a pupa similar to the butterfly caterpillar.

The Manyara tsetses do not carry sleeping sickness, for strict measures endeavour to keep all the national parks free of this scourge. To prevent infected flies being carried into a non-infected area, there are road barriers where health officers examine the cars and even look behind the sun visors.

Fortunately the governments remain fully alert to the danger and a team of scientists are constantly touring the parks where they undertake exhaustive checks and sometimes shoot the animals with drugged darts in order to take blood samples while they are unconscious. Nowadays, the infection of live-stock with nagana is one of East Africa's greatest problems.

However, there was little need for us to worry about contracting sleeping sickness, because when diagnosed in its early stages the illness is easily cured; recent drugs have been used successfully even with very advanced cases. (Later, in Uganda, we were both bitten by numerous tsetses in an area where medical surveys had proclaimed one in every thirty-five flies to be carrying the disease, but on returning to London, precautionary blood tests proved negative.) All the same, the flies were unpleasant travelling companions and we closed the car windows, preferring stuffy air to the red hot needles of tsetse bites.

The Lake Manyara Hotel is splendidly situated on high ground six miles from the National Park and there is no other accommodation in the vicinity. Way off the main road it was extremely hard to locate, for our own car tsetse fly eradication programme had delayed us those vital minutes which near the equator, where dusk is short, means the difference between driving in light or utter darkness.

On getting to the hotel we found the foyer was a scene of commotion where the manager's wife was considerably distressed. Prior to our arrival, as the sun set, a leopard had scaled the rocky cliff and snatched her pedigree boxer dog off the lawn.

Although feeding mainly on small antelopes, bush pigs and

baboons, leopards show a curious liking for dog flesh and have been known to snatch even a large Great Dane. They are excellent climbers and hunt by stealth and ambush, often dropping silently on to their prey from a tree.

When not hunting the leopard is shy and retiring, but if molested it is probably the most dangerous of animals, attacking with tremendous speed and using its fearful claws to scalp the aggressor. Powerful and courageous, only death can halt a leopard's charge, so all in all the boxer stood small chance of being saved.

The previous night the manager had suffered a harassing experience when two burst tyres stranded his van along the same harsh road where we had struggled to overtake the lorry. As he carried supplies of fresh meat for the hotel, lions and hyenas smelt the blood and prowled around the van until the assistant manager came to his rescue at 3.00 a.m.

They chased off the animals by throwing stones, transferred the meat to the car and the manager returned with his assistant. Then to cap everything their windscreen shattered as they drove into the hotel grounds! And now the boxer dog drama. For once, the gods were surely resting.

For several days, the hotel provided a comfortable respite from travel, with magnificent views across the lake and National Park. We studied the bird life, swam in the hotel pool, lazed in the sunshine and smiled at typographical menu errors which listed such appetizers as *marshy potatoes* (mashed) and *chocolate mice* (mousse). The complete relaxation made a pleasant change.

When we departed for Ngorongoro, which is among the world's greatest strongholds of wild life, it was raining hard.

CHAPTER XIII

A Wonder of the World

Pioneers – A humble diary note – Damp spirits – Gales – No gas – Soggy matches – An argument and a truce – Tragic humour – The latest rainwear – Twin fires burn – Switzerland or Africa? – A bedtime story – Steptoe looks in – Precipitous cliffs – From rain to sunshine – Reduced to a microcosm – Crushed beauty – The Garden of Eden – Jumbled memories.

ON March 18, 1892, when maps were still printing '*unexplored*' across large areas of Tanzania, a pioneering party led by the European scientist Dr Oscar Bauman discovered Ngorongoro.

In his diary, Bauman recorded the event quite simply – *At noon we suddenly found ourselves on the rim of a sheer cliff and looking down into the oblong bowl of an old crater, its bottom was grassland, alive with a tremendous number of game.*

Since that humble note was written, Ngorongoro has been acknowledged as one of the world's greatest wonders. It is the largest unbroken, unflooded volcanic caldera on this earth. The average height of the rim is 7,600 feet and with a diameter of 12 miles there is a flat floor covering 102 square miles. Over 30,000 head[1] of wild game can be viewed at close quarters from a Land Rover.

Jill and I both felt that Ngorongoro would be a highlight in our long safari and although the wet weather had remained with us all the way from Lake Manyara, our tired spirits were not daunted even when we arrived in torrential rain and mist at the Crater Lodge.

This lodge consists of individual log cabins. Initially their attractions were hardly noticed for they were spread out over

[1] Certain estimates made forty years ago quoted this figure at 75,000.

the hillside and as we had neither raincoats nor umbrellas we had to 'run for it' from the reception centre. We arrived at our cabin breathless, very bedraggled and miserably cold, for at a height of almost 8,000 feet the gale force wind was icy.

To aggravate our condition further the bottled gas heater was unlit and the matches were thoroughly damp. As the cabins have no communication system I hurried back to the reception, battling against the bitter elements and feeling suddenly in a state of dejection. Returning with another box of matches, it was only after striking about twenty of them with stiff, perished fingers that my benumbed brain realized the gas cylinder was empty.

The rain still dripped from me and I was kneeling before the cursed heater in a puddle of water. A pile of spent matches lay in a soggy black mass on the stone floor. I restrained myself from kicking the heater. Jill sat on the bed shivering with her breath condensing to white clouds and being expelled through the nostrils like jets of steam.

'Well, I can't stand this any longer. I'm sick to death of the rain and the cold and this stupid cabin.'

Unfairly, I took exception to the tone of Jill's voice and answered her crossly. 'You've come here to see the animals not to stay in a centrally heated luxury hotel,' I snapped.

'Bugger the animals,' she replied, with an equally venomous irritation.

And then we both laughed. Actually, Jill was not really swearing for when either of us say 'bugger' something it is the sign of a truce. It stems from a true story about King George V who stayed several times at Bognor Regis to recuperate from an illness.

When the King was dying, one of his peers said, 'Don't worry Sire, you'll soon be well enough to visit Bognor.'

At these futile words of condolence King George pulled himself up in the bed. 'Bugger Bognor', he replied, and then died while laughing at this apt remark.

It is one of those tragic comments which nevertheless are still amusing and we find the very thought of 'buggering Bognor' ideal for breaking down the unintentional animosity that is often generated through mere body exhaustion.

So, having settled our enmity by resorting to base words, we mutually decided that getting some hot coffe would be far wiser than worrying about new gas cylinders. To give us some protection during our muddy dash to the lounge we unhooked the plastic shower curtain and used it as a hood.

The lounge, dining room and bar are all housed under one roof in an attractive, log building heated by a central back-to-back twin fireplace.

Two enormous wood fires were burning, and after toasting for an hour or more, eating supper, drinking two pots of coffee and a large whisky we both felt better humoured. In the meantime a gas cylinder had been delivered to the cabin, and although we got saturated once again the heater was powerful enough to dry our clothes and help to keep away the chill.

With the wind roaring outside and whistling between gaps where the wall logs were joined, we could easily have been in a Swiss Alpine chalet and it was hard to believe that such weather conditions were any part of a journey through tropical Africa.

But at least I endeavoured to create the right atmosphere. Lying in bed with the blankets pulled high, I began to read out an interesting item from the current Ngorongoro Information Bulletin. Apparently, Baron Hugo van Lawick and his wife Jane Goodall (the chimpanzee expert) and two American staff had recently been invaded by lions while camping inside the crater. Baron Lawick wrote:

> 'Benjamin was in the closed kitchen tent at 7.30 p.m. and Thomas was outside cleaning a hurricane lamp. It was already quite dark. Suddenly Thomas noticed a lion creeping up on him about thirty yards away. He threw himself flat on the ground. Benjamin also saw the lion and told Thomas to get up and walk slowly back to the kitchen tent. They both watched the lion from the safety of the tent.
>
> 'Jane and I were in another tent about thirty yards away; the main door flaps were wide open. We had a large gas lamp burning and with such a brilliant light neither of us could see very far into the darkness.
>
> 'Benjamin and Thomas then watched three lions start

creeping up to Jane and me. Benjamin shouted a warning and shone his torch towards one of the lions. His torch was not very powerful and so I walked out, trying to see who had called. Benjamin yelled something else. Still not hearing, I walked out further and then saw a lion only twenty yards away coming towards me.

'I ran back, planning to make for the Land Rover with Jane. The vehicle was parked only ten yards from the tent, but unfortunately there were another two lions next to it. Quickly we let down the door flaps and I lit the gas stove, getting some paper ready to put on the fire in case the lions came nearer. We also collected some heavy kitchen utensils ready to make a noise. Then we waited. There is no window in that side of the tent and we had no torch, so all we could do was listen.

'Benjamin and Thomas put on their radio full blast, hoping the noise may scare away the lions. Suddenly we heard a yell from their tent and a tremendous clatter of pots and pans. The radio went dead and there was deathly silence. I started to open the tent flaps in order to make a final dash for the Land Rover, but both the lions were now between us and the vehicle. We called out to the staff, but got no reply. (I thought we should hear more screams if they were being mauled). Then we heard the lions ripping a tent to pieces.

'Apart from canvas being shredded, there was no other noise, and soon even that stopped and there was only silence. Suddenly we heard somebody racing to the Land Rover. I thought – felt sure, this was one of the staff (the other having been killed) followed by a lion, and I began lighting pieces of paper, planning either to throw them out or even set fire to the tent we were in.

'To my utter amazement, we heard two car doors open and slam shut! I peered out. Benjamin and Thomas were in the front seat and the lions were now outside their tent. Jane and I raced to join them.

'Then we chased the lions from the camp by hounding them in the Land Rover. They were not in the least scared and we found they were playing with us. I am sure they thought that we were now joining in with the game.'

I stopped reading. Jill was sound asleep. My arms were frozen. 'Bugger the animals,' I muttered, and put out the light.

The next morning I pulled the curtains and jumped in fright on seeing a scruffy, unkempt Steptoe character looking at me through the window. It was a shaggy-coated wildebeest. Which of us had the greatest surprise is impossible to say and for a second we both stared at each other before he galloped away.

Although the weather had not improved we were assured the sun would be shining 2,000 feet below us. So after a very early breakfast we collected a Land Rover and set off to the crater down the precipitous, muddy, one-way track hacked from the cliff walls.

After fifteen minutes of negotiating the zigzag course, squeezing a passage between dripping, dangling tropical vegetation, we broke through the rain clouds into warm sunshine. There before us lay a perfect amphitheatre. Continuing our descent, we watched black ants grow in size until they were clearly herds of wildebeest, zebra, gazelle and other plain animals. I stopped to take a photograph; looking up towards the rim, those soaring cliff walls gave me the uncomfortable sensation of being reduced to a microcosm and our Land Rover assumed the insignificance of a discarded toy.

From rain to sunshine. From dank undergrowth to flowers. We had *hoped* to get sunshine, *expected* to see animals but never dreamed of finding so many flowers, and driving across the crater floor our tyres now left an unavoidable wake of crushed beauty as they trampled us through this vast, wild garden.

Masses of blue hyacinths turned acres of land into an ocean, while clover, knee deep, produced broad wavering lines of pink tracery. Amid the green grass were scattered petunias, their white petals flapping violently in the breeze and looking so frail it was hard to believe they can actually withstand the buffeting of a gale force wind for several days. Species of the candle lily spluttered out incandescent fountains of pink blooms. Flame gladioli stood proud and erect with sword-shaped leaves, reminding me of the Masai. There were thistles glowing a deep sunrise purple, and Ox-eye daisies a brighter yellow than buttercups.

And roaming over this colourful flower carpet was a huge concentration of animals outlined against a grandeur of highland country where six volcanic peaks rose to more than 10,000 feet.

Here was an unforgettable scene where such an enchantment breathed over the countryside that it seemed a magic world all fresh and blooming from the Creator; surely this is how everything must have been 'in the very beginning'.

We spent several days exploring Ngorongoro. Each morning we left the rim at around 6.30, driving from rain into a sunny crater. We took a picnic lunch, and did not leave until just before darkness. On these daily jaunts we experienced so many thrilling and diverse scenes they quickly became no more than a jumbled collection of vivid memories without any chronological order of happening.

CHAPTER XIV

The Kill

A polygamous existence – Helpless cubs – From serenity to 'yellow perils' – Hard training – Practising a throat slit – Incompetence – A spectacular scene – Confusion – Grim undertakers – Those who wait – Warm blood and intestines – Disgrace – Predators are like people – An all night banquet – Voice of authority – Hyenas are misunderstood – Ant bears – Superstition – A sad end – Crunching bones – Wild dogs – A Garden which will never fade.

THE carnivorous or predator[1] animals are well represented in the crater at Ngorongoro and we were thrilled to observe lions at extremely close range. Because they tend to live in groups we watched them enjoying family life; being polygamous there were two lionesses associating with one male. They had five cubs which kept rolling over like kittens and indulging in friendly battles of fisticuffs or pulling father's tail. Parental control remained strict however, and immediately the games became too rough a gentle but firm slap from a large paw brought order.

Initially, a lion cub is entirely helpless, taking about six weeks even to stumble around and another seventeen months learning how to hunt. During this time the offspring are completely dependant upon their mother's teaching, and consequently wild lions produce young only every two or three years.

For mother, the habit of living as a family group has tremendous advantages. A pregant lioness is usually brought food by the other females and after the birth either an elderly 'uncle or aunt' takes up residence to guard the cubs or the lionesses themselves work out a rota for acting as nursemaid to each other's children.

[1] Flesh-eating animals preying upon others.

These particular cubs must have been around six months old because their baby coat spots were almost faded. Occasionally they looked at us with large amber eys, but apart from such cursory glances our presence caused no concern.

Even the lionesses appeared docile and serene. Yet, had I so much as unlatched the Land Rover's door, the complete family would have been transformed at once into dangerous, growling 'yellow perils', for the apparent placidity disguises a nervy, highly-strung temperament where the slightest unusual happening is often enough to provoke them.

Strangely, as long as people remain inside a car, lions rarely exhibit annoyance at being approached. Whether they fail to associate the figures in a vehicle with human beings, or whether they have learned that human beings in a vehicle are harmless, is not understood even by experts. Certainly I had caused this group no alarm by stealthily inching the Land Rover towards them, before parking quietly only a few feet away.

While watching these truly majestic animals it was hard to believe that the cubs, now playing like kittens, would soon become master hunters powerful enough to fell antelopes with a single blow of their paws. Their training is hard. At first the mother lets them practise on half-killed, small animals such as a warthog which she holds down for the cubs to learn the technique of throat-slitting. Later they are allowed to undertake a complete killing operation, although often the young lion bungles the job and gets badly knocked about, while mother growls 'tut-tuts' at his incompetence. If they are to survive, a great deal must be learned about the habits of game.

One morning a lioness made her kill just a few yards away from us and this was a gory scene as horrifying as it was spectacular.

The incident happened while we were watching zebras, and I was casually scanning the landscape with binoculars. Suddenly, an unusual pair of ears came into view. I focused more carefully and there, almost hidden by the long grass was a lioness studying the herd. Cautiously, she raised her head and sniffed, exhibiting a distinct look of ecstasy at the scent. (Zebras are the favourite food of lions.) Though appearing timid, zebras are in fact tough and strong; they can kick hard enough to break a lion's jaw, and they offer far more resistance than a wildebeest or an antelope.

We did not have to wait long for action. From the opposite direction a growling male lion appeared, running towards the zebra herd and forcing a group of them towards the hidden lioness. It was a well-planned ambush.

The small herd broke up, and as they did so the lioness chose her victim, catapulting from the undergrowth and closing with the zebra in a few bounds at great speed. Finally she gave a tremendous jump and landed squarely on the galloping animal. At this point I was unable to see what happened exactly, but generally the lioness uses her front paws to gain a stranglehold and thereby brings the animal down on its neck which is automatically broken.

Certainly, this zebra fell as if pole-axed. Then for seconds everything became obscured in a confused tumult of threshing bodies and dust.

When the dust cleared, the zebra, with its neck obviously still intact, was fighting valiantly from an unnatural, half-crouched position. But now the lioness sprang at the kicking creature and severed its throat with a single bite. Momentarily the lifeless zebra remained crouched and then, like a deflating balloon, crumbled to the ground.

The lioness padded around the 'kill' with her tail nervously twitching from the strain and excitement. Almost immediately the male appeared on the scene, while two cubs hovered expectantly behind.

Vultures circled overhead with the foreboding silence of grim undertakers. Buzzards and yellow-billed kites had taken up residence among nearby trees. From a respectful distance, a fox-like jackal was already watching with its sensitive, pointed ears raised, and not far away the hyenas would be prowling. They were all waiting for their share of the remains.

After 'sharpening' their appetites with the intestines and a little warm blood, the lions settled down to gorging at leisure some 20–50 lb. of substantial flesh. During this lengthy time the cubs are not allowed to partake in the feast, and only when the adults have finished eating can they have their meal.

Actually, right at the beginning, one of the cubs did forget his manners and warily sneaked a lick. Father stared at him threateningly as if he could not believe this impertinence, and mother promptly boxed his ears very hard. (If only the modern child experienced such discipline!) After the cub had slunk away in disgrace, we left the banquet and decided to return during late afternoon.

Through the ages, lions have justifiably been endowed with a majestic aura, and European's call them the king of beasts.[1] They are among the largest of predators and most people regard them with awe. Yet these same people who tolerate the lion often look upon other predators like the hyena and hunting dog as repugnant, vicious killers which should be exterminated.

But those who condemn all actions of predatory behaviour are overlooking the fact that the smaller carnivores actually *perpetuate* wild life. As W. O. Nazel pointed out in *Predators Are Like People:* 'The killing and eating of one creature by another is life's assurance of its own continuance. Without it there would be no check on prolific nature. It is an assurance that no one species will overrun the earth. No predator – except man – has ever successfully wiped out a species.'

Indeed, the survival and well-being of prey species depends to a large extent upon active predators, for any area of grassland, forest or plain can support only a certain number of herbivorous[2] animals in a healthy condition.

Once a region becomes overstocked the pasture deteriorates rapidly and the animals suffer from malnutrition which weakens their resistance to disease. Predacity is nature's simple way of keeping herd sizes within sensible limits for the grazing area available.

The bloody zebra scene we had just witnessed was therefore an essential part of survival, and although such 'kills' rate among the most spectacular phenomena in nature, they are also among the most widely misunderstood. Of course, conservationists recognize these facts, and even years ago controlled experiments proved that when predators were removed from

[1] In Africa this title had already been given to the elephant.
[2] Non-flesh eating animals. As opposed to the carnivores (predators).

an area deliberately, then usually, the protected prey species developed poor health and shrank in numbers.

As this same evidence has been corroborated over and over again by long research 'in the field', any indiscriminate annihilation of the predators is severely censured by the naturalists.

And so, with every minute that passes, small and large predators are killing each other in order for life to continue. This is a mode of existence which the animal world accepts as irreversible. In Ngorongoro the zebras hardly notice if one of their 5,000 strong herd is missing and no doubt they will continue to be ambushed by the same lioness in the same place in the same manner.

Nature's kingdom is not concerned with time. Yesterday has gone and tomorrow is beyond comprehension. Today means *now*, this very second, and as the animals play, run, feed, copulate and give birth, they abandon themselves completely to every irreplaceable moment, even death.

But whether the predator's victim is a moth or a zebra, the sole purpose of their slaughter is to keep themselves alive. Only man kills without a sensible reason.

In fact, after a lion's appetite is satisfied he does not hunt for several days, and during this period the lesser creatures graze closely to him in perfect safety.

A pride of six lions will account for about eight 'kills' every month which over one year averages sixteen animals per lion. As predators are relatively scarce, their total number of yearly kills represents a very insignificant percentage of the abundant herbivorous animals;[1] in many areas it is even less than one per cent.

As planned, we returned to the zebra kill during late afternoon. The lions were now lying beneath a shady bush and half-heartedly gnawing at bones with no more zeal than if they were casually sharpening their teeth.

Vultures were squatting within yards of the mutilated body, and although some of them occasionally spread their wings in anticipation of fluttering on to the offal, the next sitting was

[1] The Serengeti alone supports more than 405,000 herbivorous animals.

9. Nearing Meru's crater lake, Tanzania.

The author and Jill with Desmond Vesey-Fitzgerald pointing towards the volcano, Tanzania.

10. Miles of remote countryside through Tsavo, Kenya.

The equator at last.

scheduled for the hyenas as an all-night banquet. The vultures' turn at the feast would be the following day, and after they have picked the carcass clean, any tiny fragments which remain are finally eaten by the ants.

Thinking we had seen the end of activity I was on the point of driving away when the male lion unexpectedly stood up. He was a splendid creature with a magnificent black mane (Ngorongoro lions are famed for this feature), and for at least a minute he kept absolutely still. Here in truth was the heraldic lion, a living image of strength and power, the personified British Lion reflecting all the idealistic nobility of our National emblem. Such grandeur filled us with reverence. Then, no longer a Trafalgar Square sculpture, he looked up at the sky as if paying respect to the Creator then lowered his head to the ground and gave a full-throated roar which swept across the steppe country as a thunderous rumble.

It was a sound of indescribable magnificence, a sound that pulsated cold excitement up the spine and made goose-flesh stand out on my arms.

After sunset a lion's roar silences all the other animals. According to Franz Roedelberger's apt interpretation, it means: 'As far as my voice carries, this is my territory.'

However, this particular roar was probably either to summon others of the pride, or merely for the sheer joy of being alive and satisfied. But usually, adult lions call to one another with gentle purring throat sounds, while young cubs produce hiccuping squeaks and also miaow raucously.

There was no second call, and as if having specifically undertaken the ritual for our benefit the male immediately lay down again with the apparent intention of sleeping.

Once again, just as we were about to leave, the scene came to unexpected life with the appearance of a spotted hyena. This animal is rarely seen during the day for it dislikes the sun and lurks among deep undergrowth, in dark caves,[1] or in old

[1] Author's Note: During the last glacial era enormous *cave hyenas* hunted throughout Europe, but today the *spotted hyena* is the largest remaining species and it is normally found south of the Sahara. The rare *striped hyena*, smaller and far less aggressive, is strictly nocturnal and solitary.

ant-bear holes[1] where the young, generally two, are born with their eyes fully open.

Drably coloured, his body was reddish-grey with a black face and pale brown, almost oval spots. About the size of a large, powerful dog, he had prominent rounded ears and a short bushy tail. His back sloped abruptly to the hindquarters, and as he slouched along with an awkward hunch-backed gait his whole grotesque attitude conveyed a feeling of sly cowardice. What a horrible repulsive creature he was, the ultimate creation of ugliness.

To be fair, many people do not find the hyena's anatomy repulsive and state that any impression of *absolute ugliness* is false and superficial, being entirely due to comparing them with our illustrious dogs, to whom they are in no way related. In fact, the hyenas represent a distinctive family of their own which, during the Tertiary Period, became detached from the cat tribe. Although now a separate entity, they can be loosely classified as a link between the *cats* and the *dogs*.

This description of my first encounter with a hyena may therefore create an unfair physical impression.

On future occasions I did admit to finding in the hyena's large, deep brown eyes a strange but attractive expression of intense wistfulness like the sad eyes of an ill-treated baby. But apart from the eyes I still thought the creatures *looked* repulsive, although whether years of living in the bush would alter this opinion I cannot say.

Yet, ugly as he may be, every characteristic of the hyena's anatomy has a definite purpose. The long front legs with their ungainly paws can pin a carcass to the ground as firmly as a spiked board holds a joint for carving. The unpleasant arched jaw and thick neck muscles enable a set of fearful teeth to crunch through the heaviest bones and to suck out the marrow. Finally, the tough, coarse fur never becomes matted or contaminated even after long contact with blood and putrid flesh.

[1] The ant-bear lives in deep burrows which are sunk well past the soft earth strata where rains penetrate. Although about as big as a medium sized pig, this unusual mammal digs with the efficiency of an excavator and in minutes can vanish completely beneath the ground. Numerous animals give birth, or recover from illness, in the dry warmth and safety of disused ant-bear holes which are scattered liberally over the East African landscape.

Because of this unfortunate appearance, hyenas are possibly disliked more than any other animal. Certainly their habits are endowed with innumerable superstitions and misconceptions.

First and foremost, spotted hyenas are not merely scavengers that feed only on carrion (dead rotting flesh), but are important predators who tackle most of the herbivorous animals such as zebra and wildebeest. Although they hunt generally in large packs at night, even a solitary hyena, if hungry, becomes aggressively bold. Geologists working in the bush have recorded instances where a hyena, under cover of darkness, has actually snatched away food which was cooking over the camp fire.

It is another common belief that hyenas are hermaphrodites. This is probably because the females have external labial swellings which are unusually developed and easily mistaken for a male scrotum. Any stories relating to bisexual activities are therefore just as groundless as the superstition that all hyenas have an 'evil eye', and are ridden by spirits wishing to gorge themselves on blood.

Nor can their eerie wailing be attributed to the devil's laughter. The wail is simply a method of communication like the grunts, chuckles, growls and other sounds produced by these noisy creatures.

Again, some people claim they have watched hyenas dig up graves to get at a human corpse, but experts say this is absolute nonsense. Such a misconception may have arisen from the Masai tradition of leaving their dead for the hyenas.

So in spite of the hyena's rather objectionable appearance, much of the abuse about their existence is without good cause; and like the vultures, they clear the countryside of festering carcasses.

Nevertheless, a lion hates and despises the hyena more than any other scavenger. The vultures and marabous which assemble around his 'kill' are ignored. Of the lone jackal who may attempt to grab a share, he is reasonably tolerant. But hyenas are always driven off with utter contempt and absolute loathing.

Perhaps the lion knows instinctively that when he is very old, or sick, or crippled from wounds, the hyenas will then tear him into shreds; this is the fate of most lions. It is a sad ending for such noble beasts whose roar once meant 'as far as my voice carries, this is my kingdom'.

True to form, as soon as the spotted hyena appeared on the scene the glutted and sleepy lions, began growling, and the male rose to his feet. But the hyena, realizing that over-indulgence stimulates apathy and thereby reduces the danger of aggression, bravely risked a lightning raid upon the 'kill'.

Darting forward, he skidded to a halt, nailed the carcass with a vice-like grip beneath his powerful front legs and ripped away a piece of flesh. We watched him crunch through some bone[1] as easily as biting a cocktail *Twiglet.* Then, just as a greyhound rockets from its trap, he leapt away. The unexpected raid was over in seconds, and only as the hyena whirled past was I able to shoot a single photograph.

Had the lions not been satiated, this hyena would never have taken such a risk. There is no doubt of the intelligence of these scavengers; they show a considerable knowledge about the habits of game, and even observe the vultures to gain the exact position and state of a distant 'kill'. And their power of scent has tremendous range and accuracy.

It was an exciting episode on which to close another rewarding day.

Although there are many leopards and cheetahs in Ngorongoro, we were not lucky or skilled enough to find any of them. The only other predators we saw were a small pack of wild hunting dogs.

Their faces are similar to the striped hyena's and they have ears like the spotted hyena;[2] in size and general characteristics they can be compared with a slender, wiry Alsatian. They are, however, a unique species, and the most competent hunters of all; but as with the lion, their very act of killing helps to perpetuate other wild life.

The wild dog is almost extinct, and the Crater is one of the

[1] Inevitably the hyena is unable to avoid swallowing a great many bone fragments and as this results in their droppings being pure white, their presence in any area is always easy to detect with certainty.

[2] Years ago wild dogs were actually called *hyena dogs.*

few places where they can be found. For some reason, man has persecuted this species with particular vigour; possibly because of the wild dog's reputation for savage cruelty to wild life and their danger to people.

Yet again, most of these beliefs are based upon fallacies. And because man generally misunderstands the need for predacity, coupled with his urge to destroy, the wild dog has became a victim of our own blood lust carried out sanctimoniously under the false claim of benevolent slaughter.

Literature has also endorsed such misconceptions. Fiction writers, including Sir Arthur Conan Doyle, have written far too many chapters describing the gory death of people attacked by wild dogs.

Accordingly, many animal 'experts' have given equally misleading accounts about the intelligent co-operation the dogs use during a hunt, how they employ a relay system of taking it in turns for chasing the prey to exhaustion, how they spread out on either flank of the prey, and finally slit the victim's jugular vein.

On paper the details all sound very credible and have that popular touch of cruel excitement. I was actually given published magazine articles describing these particular hunting methods in detail and this literature, supposedly written by an authority, unfortunately appears so genuine that some authors have used the information to produce a thrilling fabrication.

Although we never saw a wild dog kill, our reading material about these animals made us extremely wary of them. But much later, in Nairobi, I came across a paper issued by the *Ngorongoro Conservation Area* which contained a special feature on wild dogs.

Having met the dedicated conservationists responsible for Ngorongoro, I am quite prepared to accept their evidence and to reproduce it here as one of the few authenticated reports of how wild dogs *really* hunt:

> 'Between 1964 and 1966 a pack of wild dogs, resident in Ngorongoro Crater, was kept under observation. During this time the actual "kill" made by the pack was witnessed on 106 different occasions.
>
> 'When hunting the dogs do tend to flank out, but usually

only the lead dog, and sometimes one or two others, are responsible for bringing down the selected animal.

'Wild dogs never hunt in relays, and it is usually the lead dog (who started the chase) that finally closes with the selected prey. The prey is run down by endurance on the part of the lead dog(s). The "kill" usually occurs within seven minutes from the start of the chase, which is not enough time for the dogs to tire and be relieved by dogs jogging along in the rear.

'As the selected prey begins to tire, the lead dog increases its speed and eventually closes with the animal. During the chase, discipline among the dogs is so rigid that they stick to the selected animal throughout the chase, even though the animal may run through groups or herds of other animals which are more accessible to the pursuing dogs.

'Selection of the prey appears to depend upon the size of the prey animal. In a wildebeest herd, calves are invariably the individuals of interest to the dogs. The usual prey selected in the Crater are Thomson's gazelle, Grant's gazelle and wildebeest calves.

'When the prey is relatively small, such as Thomson's gazelle, it is quite literally ripped to pieces by the dogs. Grant's gazelle can run considerable distances and not infrequently elude the dogs, but when finally caught, usually put up very little fight and can be thrown to the ground by one dog. Wildebeest, on the other hand, are usually very courageous creatures and generally put up a tremendous fight before they are overcome. Slitting the jugular vein – often referred to in fiction – simply does not occur.

'The dogs always attack and kill an animal by rapid disembowelment and if the prey is relatively large it is some considerable time before death comes.'

I later found another paper which stated:

'Wild dogs have rarely been known to attack man without a good reason. If anything, they are specially inquisitive about man and will follow him with noisy barks while actually jumping above the grass to get a closer look. Unfortunately, these actions are often mistaken for viciousness prior to an attack. It is a mistake that stems from ignorance.'

When the day came to move on from Ngorongoro I knew exactly how Dr Bernhard Grzimck felt when, in his famous book *Serengeti Shall Not Die*, he wrote: 'It is impossible to give a fair description of the size and beauty of the Crater, for there is nothing with which to compare it. It is one of the Wonders of the World.'

As we left for the hot Serengeti Plains, although it was raining hard my thoughts swept away the dense clouds and while recalling every moment of those days spent in the Crater I drove through memories of sunshine. For me, that Garden of Eden will never fade.

CHAPTER XV

The Serengeti

A strange grey mass – All a question of optics – Noiseless noises – The call – A lively tent – Dawn explodes – Pings and thuds – Don't look now but there's a lion behind you – Research projects – Disgrace – Coloured tags and radio collars – Drugged darts – Pimbis know their name – An eagle swoops – Four legs block the doorway – Taking a risk – Long rains on their way – Plans for Uganda – Off to Keekorok.

NEARING the famed Serengeti we were both perplexed by a mushroom-shaped black cloud which at first hovered on the horizon, and later melted into an indefinable grey mass.

After more miles, this dark mass became a distinct, jungle-covered mountain lying squarely ahead of us with the road apparently disappearing through its dense vegetation. Strangely, our hitherto accurate physical map of the area indicated only emptiness.

Another thirty minutes passed, and although the mountain still appeared just as far away, every detail of the rocky slopes and vegetation was now piercingly clear. And to complete the scene a large, palm-fringed lake had also come into view.

Here, overlooking miles of arid countryside and incongruous blue water, was an unmapped African attraction rivalling Mont St-Michel. I stopped to take photographs.

Immediately I looked through the viewfinder the colours turned sombre, and the wooded slopes trembled out of focus as if shaken by an earthquake. Momentarily I thought the camera reflex system was failing. But the camera was faultless, and we both watched dumbfounded as the mountain disappeared in shimmering vertical trickles like a painting being washed off some glass by the rain.

Then the lake disintegrated into a scattering of small pools which quickly evaporated, and once again left the road stretching as far as the eye could see across a desolate landscape.

Alas, Africa's Mont St-Michel had been a perfect mirage.

The Serengeti covers 5,700 square miles of northern Tanzania where the elevation ranges from 3,000 to 6,000 feet. We were staying at the *Seronera*, a remote Lodge situated at 5,000 feet in the central area. Our accommodation was a tent, and with the entrance flaps tied back we looked out across vast dry plains of noisy quietness.

The quietness of gazelles flitting between the trees as if floating on air. The quietness of monkeys swinging gently from branch to branch. The quietness of elephants ambling along with their featherlight steps. The quietness of vultures circling and waiting. The quietness of being watched by thousands of hidden creatures. All this activity is the noise of quietness. A noise which the body feels but cannot hear. It is the silent call of the African wilds.

Darkness came soon after our arrival, and that evening we sat around a huge camp fire enjoying the great outdoors to its fullest. But on retiring we found the creatures which belonged to this outdoors are not quite so acceptable as bed companions, and all night we were kept awake by dozens of mice continually squeaking and scurrying over us like clockwork toys.

The next morning dawn burst over the plains in an explosive red that annointed the flat topped acacia trees with fiery haloes and turned the mist into an orange haze. While dressing, I was baffled by a peculiar but repetitive sound which came from behind our tent where Goliath was parked. Ping-ping-ping. Thud-thud-thud. The pings had a bell-like quality; the thuds were dull. I went to investigate.

I walked softly around the car and found a small green bird sitting by the rear wheel. I watched and waited.

Soon the bird fluttered up to the chromium plated hub cap to look at his reflection, and in doing so, his rapidly beating wings produced the *pings*. Then, obviously believing the reflection to be a captive relative, he fell to the ground again and promptly attacked the tyre with his tiny beak, using every bit of energy he could muster. These were the *thuds*. And after a few seconds of hammering the tyre, he would first flutter up to see whether his efforts had released the prisoner before starting the attack once more.

For such a small bird the effort was tremendous, and the poor little thing's strength was failing rapidly. I tried several methods to dissuade him from continuing this noble rescue venture but none of them worked, and after a short time he always returned. Finally I removed the hub caps, and with the bird's reflection eliminated he believed a victorious liberation had been achieved; he struggled towards a near-by tree where he collapsed onto a lower branch, thoroughly exhausted.

With my rescue of the rescuer completed, we were on our way towards the central dining tent for breakfast when a male lion emerged from some long grass and began strolling *ahead* of us and *behind* another couple who were also probably going for breakfast. Although the lion was slowly gaining upon them, the couple remained quite oblivious to their following companion.

What a quandary this placed us in. If we shouted a warning, the sudden fright might cause panic and disaster. And if we gave no warning, would the lion attack? Or could I warn them diplomatically? 'I say, old boy, don't look now but you're being followed by a lion.' No, it sounded ridiculously casual and terribly British. Besides, my voice may scare the lion into sudden aggressiveness.

Perhaps the best plan was to remain quiet, to restrain any agitation, to anticipate developments and then to take any emergency action necessary. So, we continued to follow, and all five of us padded along to breakfast.

We did not wait long for developments. The lion coughed loudly and the couple turned round. There was a hysterical scream from the girl, and foolishly both of them sprinted away in opposite directions as if trying to break a world record.

The lion stopped dead, shook his head in a puzzled manner, turned around, glared at us scornfully, shook his head again and walked off with an air of disgust. After breakfast he was seen lazily stretched out along the branch of a tree,[1] and certainly not hungry. His wanderings around the lodge had obviously been a friendly, morning constitutional and he was quite peeved by such an ungracious welcome.

Later, we reported our lion encounter but apparently these occurrences are quite common. It therefore caused little interest among the rangers, although they were most interested to know if the lion had a coloured tag sticking out from one or both of its ears.

These tags are an essential part of a long term study being undertaken by the Serengeti Research Institute and directed by Dr George Schaller and Mr William Holz. The lions are tagged by shooting them with a harmless hypodermic dart, containing a powerful drug that takes effect in minutes. While the lion is unconscious a tag is clamped into its ear, for only by identifying individual animals is it possible to determine their movements, how far they follow migrating game, their longevity, and numerous other habits about which nothing positive has ever been recorded.

Other lions have been fitted with radio 'bleep' collars but this particular electronic device has not proved very satisfactory, because the collars are soon torn off during hunting.

I could give no information regarding any coloured tags, and this lack of observation put me in disgrace. For future use, the rangers then gave me a map of the area and a questionnaire asking for the *Date Seen; Location; Sex of Lion(s); Was The Tag In Right, Left Or Both Ears?; If a Plastic Strip Was It Orange, Green, Blue, Black or Silver?; If a Metal Tag Was it Brown, Red, Blue or Yellow?*, and lastly, *Number on Tag If Visible.* I was thankful to have a pair of excellent binoculars with me!

Actually, many animals are attracted into *Seronera Lodge*, possibly because it is so isolated. We met no more lions on our way to breakfast, but came within several feet of gazelle, zebra, waterbuck, wildebeest, topi and a dozen other plain animals of which the Serengeti has over 400,000.

Particularly fascinating creatures were the rock hyraxes.

[1] Hollywood, in order to save its jungle film heroes, is largely responsible for the belief that lions do not climb trees. This is quite wrong, and lions enjoy sleeping in high trees to avoid the discomfort of ants and flies. They climb fast and easily.

Greyish-brown, they are about the size of a rabbit and resemble a hybrid species which could well have evolved from guinea pigs and squirrels. But this impression is entirely false; though hard to believe, the hyrax is in fact the closest living relative of the elephant, to which it has many similar anatomical details.

There were hundreds of hyraxes around Seronera, among the boulders of a rocky hillock on the camp's western side. Every afternoon, at about 4.00 p.m., the whole community came out to sunbathe and jump around with the agility of monkeys.

The African name for the rock hyrax is *pimbi*, and for some reason, they will always answer to this name. Shout 'Rocky, rocky, rocky' and nothing happens. Yet, shout 'Pimbi, pimbi, pimbi' and the hyraxes come bounding towards you just as a dog comes when it is called.

At first we thought these particular pimbis had become partly tame and conditioned to recognizing the name, but this was not so. We found other pimbis, far away from camps or national parks, where the characteristic was just as developed. I can offer no explanation apart from endorsing that any similar word tones like *wimpy*, *dimby* or *limby* brought no response.

Jill became fond of these pimbis, and by feeding them regularly upon vegetable scraps from the kitchen, they soon became tame enough to recognize her, and would come running after us from a quarter of a mile away.

Apart from boa snakes and leopards, the hyrax's main enemy is the eagle and Jill was very upset when during the early evening one of these fierce birds fell onto a pimbi while he was eating a cabbage leaf. It was yet another example of predacity carried out with speed and judgement.

After the lion episode, our next comparable animal experience was Jill's adventure.

We had moved our accommodation from the tent, because we discovered that hundreds of mice were breeding in holes beneath the ground sheet; we were now in a small brick-built room with the toilets and washing facilities some yards away.

Jill was suffering with dysentery, caused no doubt by the

camp's practice of refreezing left-over food and then serving it hot a day or so later. On this particular evening I had taken two Sonalgin tablets (these contain a narcotic alkaloid plus a barbiturate for inducing sleep) to help relieve the pain in my ankle which I had sprained that afternoon, while climbing some rocky slopes for photographs.

Although I remained fast asleep, Jill's dysentery was necessitating frequent visits to the lavatory. At 2.00 a.m. she tried to wake me up.

'There's a giraffe standing right outside the door and I can't get past,' she informed me.

'Shoo him away then,' I answered unhelpfully, as the Sonalgin put me back into narcosis.

The next morning Jill told me the entire story. She had opened the door to find four legs completely blocking her way. A giraffe was eating the leaves of a tree which spread its branches over our roof.

Jill banged on the door and dutifully 'shoo'd', but the giraffe still did not move. Then, taking the torch, she cautiously shone the light upwards, but the giraffe merely looked down at her. Finally, after all attempts had failed to move him, Jill simply gave his legs a gentle push while keeping inside the doorway for some protection should he kick. It was a dangerous action (she said his legs felt as thick and as tough as a metal pipe) but at least the giraffe shifted enough for her to squeeze by.

As the months of long rains were fast approaching, our stay in the Serengeti was relatively short. and we thought it wise to press on towards Nairobi. We could then leave Kenya early enough to have at least six weeks in Uganda before floods prevented any travel to the extreme south (on the Congo and Rwanda borders) and the extreme north which is on the Sudan border.

We were going to break the journey to Nairobi at Keekorok.

CHAPTER XVI

Respite in Civilization

Sheer luxury – An unusual surgery – Venereal disease – Borrow my wife – Snakes for handbags – Short cut to disaster – Buffaloes charge – Dented and dazed – Masai commercial progress – Approaching Nairobi – Where safari life begins – Love or hate – Lift service – Takeover – Night beauty – Personalities – Independence – Rapid maturity – Misreporting – Mob violence solved – Watching the world go by – Photographic fascination.

FROM Seronera it was about sixty miles to the Kenyan border and Keekorok Lodge, where we were breaking our journey to Nairobi. Owned by Block Hotels,[1] this lodge had an attractive swimming pool as well as every comfort and facility. There was no question of 'roughing' it here; this was sheer luxury.

While exploring one of the several villages around Keekorok we came into conversation with a young boy who introduced himself as 'Doctor' although he could have been no older than eighteen. At first I imagined his title must originate simply from wishful thinking, but he dispelled this notion by insisting that we should visit his 'practice'.

He took us to a round, thatched hut where 'surgery' was chalked on the door in capital letters. Proudly he asked us to enter. It was one large room, very clean and smelling antiseptically sterile. There were fitted wall cupboards, chairs, a refrigerator (no doubt for preserving certain drugs and snake serums that are sensitive to heat) and two tables.

On the centre table stood the handwritten sign 'Doctor',

[1] The Block organization which operates from Nairobi, is a household name for comfort and service and has some of the best sites in East Africa. Jack Block is an enthusiastic photographer and his shots of wild life are the envy of many a professional.

while on the side table another sign read 'Dispensary'. The latter table was partly covered with bottles, tins and equipment.

The poison cupboard had a padlock and was clearly marked with a skull & cross-bones in red chalk; somewhat unorthodox but certainly effective! Everything appeared neat and orderly. I have seen far less respectable surgeries in Europe.

Having expressed our genuine admiration I then asked the doctor what diseases were most prevalent.

'Malaria, anthrax[1] and a few others. See for yourself.' He thrust a 'Patients Book' into my hand.

The columns were headed *Full Name*, *Address*, *Symptoms*, *Diagnosis* and *Treatment*. To have such personal information offered so casually made me feel rather embarrassed, especially when I glanced through the pages and saw only isolated entries for malaria, bronchitis, rheumatism, tuberculosis, typhoid fever and anthrax, but dozens upon dozens of gonorrhoea, with a few cases of syphilis thrown in for good measure. I supposed these last two diagnoses came under the doctor's category of 'a few others'.

Apparently venereal disease is a serious problem among the Masai and, because it causes infertility, is another reason for their declining numbers.

Years ago venereal disease was unknown among the Masai, but those who live close to developing cities such as Nairobi, have experienced a great influence on their way of life. Consequently they realize money is far more important than cattle.

The Masai women are often extremely beautiful, and have a natural ability for pleasing a man sexually. They quickly found it was very easy to earn money by 'sleeping around' with tourists and businessmen.

With no knowledge of prophylactic measures they soon contracted disease. This spread through their villages at an incredible rate, due partly to the Masai tradition of honouring visiting male friends and distant relatives by providing one of their wives as a bed companion. The doctor told us that Masai who live close to the larger towns are rarely free of gonorrhoea for more than a few days after being cured.

[1] Anthrax often strikes the Masai cattle and because the herds are so vast the infection is easily contracted.

I was most impressed with this young doctor, and when he asked me for the address of any British educational institute which ran overseas postal study courses for the Oxford General Certificate of Education it made me realize just how much this boy had achieved by his own determined efforts. He claimed no academic qualifications, yet had taught himself enough about medicine to be appointed officially as an 'unqualified doctor'; now he was saving lives, giving injections, diagnosing by auscultation, mixing medicine and prescribing drugs. It was a great achievement.

Walking back from the surgery we passed a village shoemaker who was cutting up a beautifully marked snake skin, for a lady's handbag. In Swahili he told us the 25-foot boa snake had been speared by the Masai. These snake skins are about eighteen inches wide and the Masai manage to kill three or four boas a week. The shoemaker buys the snakes and sells the handbags he makes from them to a shop in Nairobi. He also makes slip-on sandals from old car tyres but these alone do not bring in sufficient money to feed and clothe his family. Originally, his trade was producing spears for the Masai, who paid for them with milk and cattle. 'Those were the happy living years,' he concluded sadly.

After leaving the shoemaker, we cut across a tract of scrubland to avoid the unshaded walk along the dusty twisting road to Keekorok. Almost half-way there, I was surprised to see a Land Rover speeding in our direction. It was driven by an armed and irate game warden, who made room for us on the front seat with him.

He had spotted us strolling through an area full of buffaloes with their calves. When their babies are very young, a 'mummy' will tolerate no intrusion and to meet her on foot would hardly stimulate a friendly acquaintanceship. Of course, we ought to have known better, but on seeing Keekorok so close, my usual sense of caution had temporarily deserted me.

The warden had not exaggerated the danger and the next day, while in a hired Land Rover exploring the rolling grasslands of *Masai Mara*, a small herd of buffalo with babies took umbrage at our presence. They began a tentative charge,

1. Picnic on the car bonnet in Zanzibar.

Crocodiles blend into the landscape. Along the Nile, Uganda.

12. Arnold Klosser and Jill load up our plane, Mafia Island.

Buffalo are always inquisitive, Keekorok, Kenya.

splitting into an evenly divided 'V' formation with the apparent intention of undertaking an attack from both sides. But there was no real animosity and, having proved themselves to be made of sterner stuff than mere cows, they were content to watch our Land Rover race away without giving chase.

We had been very alarmed for a few seconds, despite the solidity of our Land Rover. These vehicles, produced by Rover of Birmingham, have a tremendous reputation throughout East Africa and are an excellent ambassador for British engineered products. Although Japanese, Continental and American models of a similar design are also available, spare parts are difficult to obtain; and in many instances their ability to withstand the toughest conditions has proved unsatisfactory. For most work, the Land Rover is still regarded as unbeatable.

Funnily enough, Rover's current advertising campaign was specially topical because it showed the picture of a Land Rover being charged by a buffalo. The copy heading ran: *The Result? One Very Slightly Dented Land Rover. One Very Badly Dazed Buffalo.* But whether a 'V' formation of about twelve buffaloes would produce a similar result I would hardly like to find out. The thought of *twelve* badly dazed animals may be stretching things a little too far even for a British Rover!

Before leaving Keekorok we visited another Masai community, but lying on a tourist route what a different welcome we received. Friendly, yes; but utterly spoiled by visitors and an American film company which had been out there quite recently.

The chief wanted 20 shillings[1] for us to enter the village and another 2,000 shillings if we took any photographs. Admittedly everyone was rapidly dressing themselves in some of the most colourful and intricate beaded jewellery we had ever seen among the Masai, but 2,000 shillings was a ridiculous price. The final disillusionment came when a girl handed us her exquisite arm bracelet to examine, and on the clasp it said *Made in Japan*. Such is progress!

[1] The basic unit of currency is the East African shilling (abbreviated Shs.) Although in line with the British monetary system (Shs. 20 equals £1) the currency is decimalized with the shilling being divided into 100 cents.

Although declining to enter the village we distributed sweets to the children and everybody waved goodbye with great friendliness. For at least a mile the swarms of Masai flies stayed around us, and when these had gone the infernal blood-seeking tsetses took their place.

Most of the road surface to Nairobi (160 miles) was absolute hell, but impressive views of the Great Rift Valley more than compensated for the continual bumps and general discomfort.

Stretching from the Red Sea through East and Central Africa, the Great Rift Valley cuts through Kenya from north to south and carries on to Malawi.

This valley is a gigantic earth fissure which dates back some 15 million years to when the earth's crust was split assunder by a terrestrial upheaval of the first magnitude.

In Kenya, parts of the Great Rift Valley are 3,000 feet deep and forty miles wide and such dimensions give an indication of the tremendous power unleashed at that time. We found one particularly impressive scene just twenty-six miles from Nairobi where the twisting road skirted forest-covered slopes which dropped sheerly for almost 2,000 feet to the plains below.

Nairobi itself is a well-planned, modern capital situated at an altitude of 5,452 feet. The outskirts are very attractive and, on an excellent road, we passed fields of cultivated pyrethrum,[1] coffee and tea plantations, large tropically designed houses, roadside markets, the Banana Hill Shopping Centre, open-air cinemas, numerous public flower gardens and small parks.

Only sixty years ago the site of this capital was the border between Masailand and the Kikuyu territory, where the big five[2] roamed. As little as ten years past, the hyena's cackle was still a noise frequently heard in Nairobi at night. Even now, this city is close to the pulse of East Africa's wild life and people living in the outlying residential areas often have poultry yards raided by hyenas, while very occasionally the scent of their dogs will attract a leopard.

[1] A type of white flowering chrysanthemum from which the ingredient *pyrethrum* is extracted; this is used in many 'aerosol' insecticides.

[2] Elephant, buffalo, rhino, hippo, lion; commonly termed *The Big Five.*

Such facts were hard to realize as we turned into the wide Kenyatta Avenue, where four traffic lanes were divided by voluminous green trees and lined with parking meters.

After these months of travel, the initial fortnight I had spent in Nairobi on my own seemed far away. It felt good to be back. The sunshine, the avenues, the white buildings, the clear air and the colourfully dressed people all offered their individual welcome. It was like returning home.

At the end of Kenyatta Avenue is the New Stanely Hotel, which has been world famous for almost twenty years. Indeed, many best-selling novels and travel books about hunting expeditions have opened with a description of the ever-changing crowds in this hotel's foyer, the meeting place for every type of adventurer and tourist who visits East Africa.

Safari life begins at Nairobi, and the New Stanley is the hub of it all. The Stanley's guest book reads like a *Who's Who*; even during my own fortnight there I rubbed shoulders with the Duke of Bedford, William Holden and Armand and Michaela Denis.

Since 1949 the Stanley had been managed by Brian Burrows, and its success was very largely due to this forty-six year old Britisher, who studied hotel management in France, Switzerland and America. When I had the opportunity of talking with Brian it was surprising to learn that 75 per cent of his staff are still illiterate, and only forty years ago their fathers dressed in animal skins.

Literate or not, every one of them gave a faultless, attentive, friendly service, and having been trained exclusively by Brian they now regarded him as some kind of Dutch uncle.

During our conversation, Brian had told me: 'You either love the Africans or you hate them, I love them. And Kenya itself is a wonderful country. Here is my home and I will never want to live elsewhere.' There is no need for him to say more. These short sentences revealed the solid foundation on which Brian had built the Stanley's reputation, for a genuine love of the country and its people is the strongest bond any hotel manager can forge between his staff and his guests.

I had enjoyed my stay at the Stanley, and often I had sat in the foyer for hours on end watching the safari crowds come and go. Luxury hunters, real hunters, tourist hunters. Novices, experts, pseudo professionals, true professionals. Naturalists. Wild life photographers. Animal sightseeing tourists. After a time it was possible to categorize them all quite accurately from their general talk, behaviour and dress.

I even managed to detect a similarity in the manner various nationalities pushed the lift button! The British often exhibited an air of disbelief: 'Of course, we can't expect the lift to work with bloody Africans maintaining it,' they were thinking. Germans strode across the foyer like a conquering army and pushed the button hard and firmly as if summoning their personal servant for an immediate reprimand. The ever cautious Swiss pressed it with suspicion in case African lifts were not a free service; to save a few cents they would rather climb the stairs. Americans had their finger outstretched from at least thirty feet away; time was vital. The French caressed the button. The Italians did not even see the button and just waited.

But the New Stanley, like everything else in the world, is wide open to sudden development and unexpected change.

When we arrived back, this hotel had been taken over by Hilton International, and in accordance with their policy Brian Burrows' managerial chair was occupied by a Hilton-trained man from Europe.

Although Hilton have planned big renovations to inject more local atmosphere into the rooms by making extensive use of native craftsmanship, and although their financial resources will enable the hotel to be continually refurbished with every facility, there are hundreds of regular guests who will never feel the same affection for the *Stanley Hilton*[1]. I am one of them.

Yet, I fully realize that, as tourism increases, only organizations like Hilton have the resources and administrative know-how to keep abreast with our changing times, and international hotel chains of this calibre will offer increased accommodation and attract a higher percentage of yearly visitors to Nairobi. More people can therefore enjoy the thrill of a safari into wild Africa, and this can only help to inspire a new interest and an awareness of nature's rich offerings to mankind.

[1] The *Nairobi Hilton* is also under construction.

Apart from the sad absence of Brian, the faces of the staff were still familiar, except that a beautiful African receptionist who had been at the 'stamp and parcel' counter was missing. 'Sacked,' a reliable hotel employee told me. Unfortunately, she was using her position to note the single men arriving. One night, at 2.00 a.m., this beauty became rather daring and borrowed the pass key to a man's room she had visited already on several occasions. Having opened the door silently, she slipped off her dress in the darkness and completely naked lay down on the bed next to the man's wife (who had joined him that day) and whispered: 'Darling, I am here again.'

She had never knocked on my door!

Kenya was the last country in East Africa to become independent, and consequently she benefited from having seen the initial administration mistakes made by Uganda and Tanzania. As this book is not a political treatise I am omitting any reference to the Mau Mau[1] terrorist activities. It is sufficient to say that modern Kenya is the most developed country in East Africa and President Mzee Jomo Kenyatta has proved to be an impressive, highly capable statesman and a good friend to the West.

The much publicized national policy of 'Africa for the Africans' has never meant kick out the Europeans regardless, but instead, Europeans have been given an incentive to stay and to gradually prepare the Africans for taking over. In this direction Uganda and Tanzania made a big error, and by forcing the white people to leave *en masse* the progress and development of both these countries ground to a standstill. Now they pursue, below the surface, a flexible policy very much the same as Kenya's, but recovery from such an upheaval is naturally slow.

(In Kenya, far more discrimination is being shown against the Indian population who are being squeezed out of the country as non-citizens by a government which *rightly* considers

[1] On October 20, 1952, as a result of widespread murders and other terrorist activities by the illegal Mau Mau organization, a State of Emergency was declared by the government. Not until November 1956 was the army able to withdraw from active operations. By July 1957 conditions in the Colony had returned to normal and Independence was granted on December 12, 1963.

that its first duty is to its own citizens, who, by virtue of an Indian exodus may be given additional employment and further opportunities. Tanzania and Uganda plan to instigate similar policies.

It is said that apart from exploiting the African, few Indians have ever done anything but damage the economy and political stability of Kenya. Personally, I have every sympathy with President Kenyatta's views, and I am only sorry that so many who fled from Kenya were able to arrive in Britain with no British ties, no British allegiance and no connection whatsoever with Britain except an incongruous British passport which allows them to live on our Welfare State and to be supported by the British nation they openly claim to despise.)

Of course, Kenya also showed immaturity in many ways, and immediately following Independence all native dances (especially tourist exhibitions) were forbidden as savage and primitive. However, this ban lasted for less than one year before the government realized that such dances were neither savage nor primitive but an important part of African culture and history. Kenya was growing up rapidly.

Speaking generally, I found that East Africa (Kenya and Uganda in particular) appear to be the most settled of any newly independent countries I have visited. The Africans are intelligent and humane people. To a very large extent they have shaken off that ugly 'chip on the shoulder' which mars so many dark-skinned races including some of their neighbours in West and Central Africa.

Political correspondents may write a great deal of derogation about the conditions of 'black' versus 'white' in East Africa, but often a situation is very different from how it appears after a twenty-four hour flying visit. A well-known example was the political hack who wrote *African Tribe Takes Up Arms Against Whites*; the information was gleaned from two excitable visitors and the tribe in question comprised six bow and arrow armed warriors waylaying tourists to give them the opportunity of taking photographs for 25 cents.

Likewise, due to the enormousness of Africa, the public tend to misconstrue even accurate news reports and Kenya experienced thousands of holiday cancellations when trouble in the

Congo was at its height. This was rather like deciding not to visit Amsterdam because there were riots in Naples.

Of all the Europeans I talked with in East Africa, about 90 per cent had only kind things to say about the Africans. Without arguing the rights and wrongs of colonialism, let no nation sneer at the so-called British colonialist types who have remained in East Africa. These British people are performing a damned good job. There are school teachers, doctors, farmers, technicians, animal conservationists and hundreds of other skilled men and women dedicating their lives to help the Africans; many of them work under tough conditions far from civilization.

Gratifyingly, I doubt if any race is more worthy of this help, and at least the majority of Africans appear to appreciate everything being done for them.

While it is natural for any reminder of colonialism to be repugnant, British law, education and technology is still regarded as the finest in the world, and a rich inheritance. As a high ranking government official admitted: 'We have a great deal to thank Britain for, even though we did not realize this at the time, and even though we often fail to admit it now.'

Throughout our long travels we developed a close bond with the Africans at every level, and where I have every objection to admitting thousands of Indians into Britain, I have no compunction about opening our doors to the African.

The Africans are very different. Usually their sole object in getting to London is to undertake an educational course and they have no intention of remaining here once it has been completed. I have never visited any nation where the people are so conscious of the need for education. And since Independence, Africans have shown a tremendous pride in their country with the creation of fine new buildings, schools, universities, hospitals, public parks and much stricter rules against national park poaching.

Where there are power struggles in East Africa, they mostly originate from the top, and ordinary people cannot be concerned with riots and demonstrations simply because the population is too widespread. Whereas 100,000 Indians could be on the streets of Calcutta in minutes, it would take days to collect (by bus) so many Africans together. Lack of communication is therefore an apparently happy solution to mob violence!

Being truly cosmopolitan, Nairobi attracts strange personalities from all over the world and there is an abundance of those who offer unusual 'therapeutic' services for the health of mind and body. Some are quacks and some are genuine. Almost certainly they have been drawn to Africa because it has always been the mysterious continent from where rumours have constantly originated about witchdoctors and remote jungle communities who have the secret of eternal life, or a cure for cancer or some other illness which has so far defeated civilization.

Today, such rumours are equally prevalent and from time to time they are still given news prominence by our more sensational newspapers.

In addition to physiognomists, palmists, psychics, physiologists and physiotherapists, Nairobi happens to be the only African town where a *physaromatherapist* may be consulted.

The title is 'coined', but the art of *aromatherapy* was practised by the Chinese 10,000 years before Christ was born. It is a special type of healing whereby herbal oils are massaged into certain skin areas that are particularly sensitive to absorption. A French doctor is known to have arrested gangrene by using this method during World War I, and those who practise the therapy believe there is a herbal oil for every kind of body or mental trouble.

The physaromatherapist in Nairobi is Miss Faith (an unfortunate name perhaps), and her knowledge of this ancient subject is fascinating. For example, I would never have known there are 689 different rose oil extracts, all of which have a distinct use, from healing wounds to producing anaesthesia or improving the memory.

Lemon grass oil is a tonic and an energy restorer, violet is a tranquilizer, orange[1] is a stimulant, sage helps rheumatism, geranium improves certain sinus and blood circulation ailments, sandalwood benefits vascular problems and cactus is used for skin rejuvenation.

[1] To manufacture 1 lb. of this orange extract takes approximately 4½ million blooms. And for rose oil, 10 cwt. of petals are required for even less than 1 lb.

So, if someone is having a rheumatic pain in the shoulder, or feels a bit 'off colour', then a visit to Miss Faith can hardly do any harm. At least she will impart a great many facts about this particularly interesting subject . . . she also happens to be a very beautiful young lady!

But to wander around the shops, to visit the market, to sit in the public gardens and to drink tea or coffee at an outdoor café is perhaps the best way of feeling the atmosphere of Nairobi and getting to know the kind of people who belong to this modern African capital.

Businessmen will nod a friendly greeting. Someone else, with genuine interest, will ask if you are on holiday and if you like Kenya. A beggar with mutilated arms and legs holds out his hands; you feel in your pocket and find no loose coins. 'I'm sorry.' The beggar grins. 'Never mind, bwana, perhaps next time,' he says. And the next time, as little as 10 cents (1d) will bring grateful smiles and a 'Thank you, thank you, bwana.' (How different from the surly, threatening, beggars that had pestered us throughout India.) Fat mammas waddle along with a happy smile while humming a tune. Young girls pass by: office workers, shop assistants, television[1] personalities and those who are simply looking for a lone male tourist. Some are dressed in mini skirts, others in well-cut European clothes and a few in African styles. But most of them are attractive, and their eyes smile as you admire their peach-smooth complexions, their well formed breasts and their long, slender legs. Warm brown bodies full of sunshine and love and simplicity and laughter and sensual pleasures.

Whether the men wear dark suits, tribal attire or jazzy shirts, their movements exhibit the slow dignity indicative of an African.

Young confidence tricksters, endeavouring to appear like university students, sport a pair of spectacles with plain lenses,

[1] VOK Television (The Voice of Kenya) was launched in October 1962, and although the transmission was intended to cover a radius of only 25 miles from the centre of Nairobi, excellent reception was received 90 miles away. In 1964 another transmitter was built to serve Western Kenya and a further station is planned at Mombasa. All programmes incorporate advertising commercials and the total transmission output is 38 hours per week with 15 hours of live performances. There are approximately 11,000 receivers in operation.

realizing that even the lowliest African looks a scholarly type when he wears gilt framed glasses! These shams hang around the New Stanley. 'Excuse me sir, I came to Nairobi for the day with a party of horticultural students and I seem to have missed my return coach. Could you possibly let me have five shillings for a meal? I do feel terrible in having to ask.'

Some nights later, you are stopped by the same 'student'. 'Excuse me sir, I came to Nairobi for the day with a party of biologists. . . .'

You cut his rigmarole short, 'Last time you told me you came with a party of horticulturalists.'

He looks sheepish. 'If you gave me five shillings then I'll give it back now,' says this unprofessional 'con-man' with a disarming smile. And he would be quite willing to do so.

To watch the succession of clients visiting a Photo-Me[1] automatic studio is yet another entertainment only equalled by a seat in the New Stanley's foyer.

The Africans are fascinated by any kind of ingenious scientific gadget. A machine which automatically takes photographs and delivers four quality prints is something no less than real magic to them. All day long the Photo-Me Studio kiosks are kept busy by pretty girls making ugly faces, ugly girls trying to make pretty faces, girls who obviously wish to look alluring, jovial men who adopt the seriousness of an undertaker, serious men who attempt clownish joviality, fat women, thin men, children, mothers with babies, and even Masai warriors who always try to get their spear inside the kiosk but eventually leave it outside.

This is Nairobi, and one can feel only affection towards her varied people. Children, fat mammas, beggars or Photo-Me fiends. Businessmen or 'con-men'. Good girls or bad girls. It makes no difference.

[1] *Author's Note:* In England, Photo-Me Studios are situated in shops, arcades and railway stations. For obtaining Passport and International Driving Permit photographs they offer a big money saving by delivering four excellent photographs for 2s 6d.

CHAPTER XVII

Treetops

A lofty perch – The brain wave – Built in an unexplored valley – Stage management to perfection – An historical plaque – Mau Mau activities – Preparations for battle – Victory – Discipline – Enemies – Familiarity breeds contempt – The Unloved Ones – Almost extinct – Monsters – Nostalgic dancing – Grasshopper thoughts – Death is kinder – Life is complex – Threats and compromise – Strange adoption – Clever strategy – Easy targets – Atrocities in perspective – What Prince Philip saw – Peter Scott has the final word.

NOWHERE in the world is there a more unusual and fascinating hotel than Treetops. Situated at an altitude of 7,000 feet in the Aberdare Mountains, a few miles south of the equator, Treetops, although surrounded by forest, overlooks an enchanting pool in a natural clearing. As the name implies, this wooden building is perched loftily in the top of trees thirty feet high. Animals accept the presence of Treetops without suspicion, and there can be no finer location for studying wild life in complete safety.

The origin of the hotel dates back to 1932 and was entirely due to the brain wave of British-born Eric Sherbrooke Walker who evolved the idea after an uncomfortable experience with a tourist and a rhino. In his best-selling book *Treetops Hotel* (published by Robert Hale) Eric Sherbrooke Walker describes with great humour that episode which produced the flash of inspiration responsible for Treetops.

'For two exhausting hours we crawled through the African bush, the plump American stockbroker and I. The wait-a-bit thorn clutched at our clothes. Tropical sun beat down on our heads. Holding my heavy rifle, I cautiously led the way to a clearing where I thought a rhino might be standing in the shade of the trees. I studied the wind, and birds which could give the alarm. I looked for forest animals which would dart off noisily at the sound of our approach.

'At last, there he stood – a massive two-ton rhino. He was in the clearing, facing us, absolutely motionless except for irritable flicks of his ears. His horn pointed menacingly towards the sky.

'The stockbroker panted up behind me. I signalled him to keep crouching behind the tall grass. The beast might hear or smell us, and charge. It was a tense moment. We had no cover nearby except light bush. But here was the exhilaration of success; the achievement of bringing a clumsy city-bred human within a few yards of one of the most dangerous animals in Africa.

'Suddenly my companion stood up. In full view of the rhino he pulled out a pocket notebook and fountain pen. Then he crossed out the word *Rhinoceros*.

'Now show me an elephant!' he said loudly.

'Fortunately the rhino took not the slightest notice, but I did. I resolved never again to stalk animals in the bush with unappreciative strangers. What was the alternative? Every year, hundreds of people came to my hotel, the Outspan at Nyeri in Kenya, and almost without exception they wanted the thrill of seeing big game. Like the stockbroker, who was more at home in the jungle of Wall Street than the wilds of Africa, there were sometimes among them rather heavy footed townspeople with little knowledge of bushcraft and stalking. A few were too well fed or not active enough for climbing trees if the animals turned troublesome.

'It made me think of a story I had heard as a boy, about a Frenchman who wanted to learn monkey language. He had a great iron cage made, put it beside a jungle stream, filled it with food and stayed there in perfect safety with wild animals all around him. Next I thought of Peter Pan's house in the tree, and the Swiss Family Robinson with their tree home, *Falconhurst*, reached by a ladder inside the hollow trunk.

'Here, surely, in these stories of the fascination of tree houses, was the solution I needed; we would build a branch hotel –

literally – of the Outspan, fixed in a tree, where our guests could safely see wild animals at close quarters in natural surroundings, and the animals would not be inconvenienced by the guests.'

And so Treetops was born and it was built in a forty mile-wide valley overlooked on one side by the snowy peak of Mount Kenya and on the opposite side by the Aberdare Mountains, where elephant, rhino and buffalo roamed. Apart from a few hardy pioneers who settled in the valley for a short period just before the 1914–1918 war, its forests, plains, rivers and trout streams had never been inhabited or even fully explored.

As it is not possible to arrive at Treetops unescorted we booked a UTC tour, and with thirty-one other sightseers we were transported the 120 miles from Nairobi in five of UTC's black and white zebra striped Volkswagen buses.

Just before midday we arrived at the cool, mountain town of Nyeri. At the Outspan Hotel an excellent lunch had been prepared. (The founder of the Scout Movement, Lord Baden-Powell, spent the last years of his life at Nyeri. 'The nearer to Nyeri, the nearer to bliss,' he had often remarked).

Later, after leaving behind our heavy baggage,[1] we set off on the short run to Treetops in a specially converted Ford Transit van, which took fifteen passengers and then returned to the Outspan for its final load. The group is deliberately split because it is difficult for Treetops' narrow staircases to handle all the guests efficiently if they arrive at the same time. Furthermore, a larger group is much more difficult to protect from any encounter with animals.

Even before Treetops had come into view the road petered out to where a white hunter, complete with rifle, awaited our arrival. He addressed us in almost whispered tones.

'Now, I want you to follow me. Stick close together and don't talk. If a buffalo or a rhino makes an appearance dive into one of the protective hides – they're not gentlemen's

[1] Although of solid construction, any unnecessary weight at Treetops is naturally undesirable (apart from the inconvenience of hauling heavy suitcases up the narrow wooden ladder type stairways) and guests are requested to take only an overnight bag.

toilets – which you will see built at intervals on both sides of the track.'

The white hunter may just as well have saved his breath, for we set off in a noisy scattering of people who stopped continually to take photographs.

'Buffaloes and rhinos. What nonsense! Just a little stage management to give some atmosphere,' said a fat, loud-mouthed English-speaking German tourist.

I secretly wished a rhino would charge out of the bushes and make him run, but nothing so dramatic happened. However, when the second group arrived some forty minutes afterwards, a snorting buffalo did in fact take exception to the chattering voices, and from the roof terrace we watched everyone race to the protective hides where they were marooned for the next fifteen minutes.

I was gratified to see our fat German watching the escapade, and having once lived just outside Cologne I could not resist making a comment to him in German. 'That stage manager must be really first class,' I said, while maintaining a serious expression.

Our bedroom (Number 3), was ideally positioned with a full length window overlooking the water hole. On consulting the visitors book, we found that almost exactly a year previously, Princess Marina had occupied it.

As would be expected, the accommodation was extremely compact, but there are comfortable beds and both stories have washing facilities at the end of their short, narrow corridor. Single guests must share a bedroom.

There is a small attractive bar and an oblong dining area where tree trunks protrude out of the floor and disappear through the ceiling!

The roof of Treetops has ample space for all, and the comings and goings of the animals can be seen on its four sides. The sunshine gleams on a commemorative plaque.

In This Mgumu Tree
Her Royal Highness the Princess Elizabeth
AND
His Royal Highness the Duke of Edinburgh

IV. Buffaloes along the Nile, Uganda.

Spent the Night of February 5th 1952.
While Here Princess Elizabeth
Succeeded to the Throne Through
The Death of Her Father
King George the Sixth

Below the terrace roof, jutting over the water's mud wallows, is an open verandah fitted with a line of comfortable aircraft chairs donated by a friendly airline. Visitors can sit outside at night and gaze down upon the animals, who apparently regard the artificial illumination as moonlight.

In 1932 Treetops hardly offered so many conveniences. Water had to be delivered in buckets and hoisted up with a rope pulley system. Electricity was supplied by banks of car batteries, and a single Elsan was reached by a hair raising foot-walk along an overhanging branch.

Many improvements came about after Treetops was burned down by the Mau Mau in 1954. The new building, once again inspired by Eric Sherbrooke Walker, is more congenial and functional. (And the afternoon light now shines from *behind* the water hole, whereas previously, after 3.00 p.m., observers were facing directly into the sun which made photography almost impossible.)

In spite of the additional comforts, the charm of this new Treetops, according to the veterans, has remained unchanged.

We were upon the roof that afternoon when the manageress (the wife of the Outspan's manager) asked excitedly if she could borrow our binoculars.

She told us that the scores of baboons now playing around Treetops were intruders who had moved in three days ago while the resident troop were away on a routine jaunt to the deeper forests.

Now, she had spotted her old friends arriving back. It was an interesting situation.

The intruders became quiet, sitting around expectantly as their field marshals began relegating the males into battle positions. The women and children were sent away to hide in the undergrowth. The home-comers were also preparing themselves, and a subversive infiltration was already under way for a surprise attack on the side flanks.

A great calm had fallen upon the animals. Even the birds were hardly twittering. Then a piercing scream broke the stillness and in a trice everything was bedlam. Screams, barks, grunts; a whirl of threshing bodies, tree-swinging fights, showers of leaves, the crack of branches, turmoil and dust.

The intruders, although badly shaken and disorganized, managed to hold their ground. The rightful owners retreated, and regrouped for a second swift attack, which followed within minutes.

This time victory for the rightful owners was decisive, and after only brief, spasmodic clashes there was a general exodus of the intruders to the hills. Immediately the troop leader, an enormous specimen about four feet tall who our hostess called by name, climbed up to the roof. Ignoring the spectators he stood on the extreme edge, jumping up and down in excitement, thumping his chest and uttering loud shrieks of conquest.

In any Tarzan film, such an episode would make the adults give superior titters at the naivety of Hollywood, yet here in 'natural vision' was a coloured, three-dimensional Cinema-Scope drama captured in the media of real life.

Every action of the baboons had obviously been carried out with planned strategy. Baboons are highly intelligent, even though their frolics and capers may not give this impression; they are among the most strictly disciplined creatures of the animal world, and we had just seen this fact clearly demonstrated.

Like many humans, baboons are afraid of the dark. Just before sunset their group leader climbed into a tree and uttered a yodelling 'ow-oo-ow-oo'; the signal to retire for the night.

Immediately, all the gambolling antics stopped, and the baboons set off in columns towards the forest, where they sleep at the top of tall trees. Small babies[1] were carried beneath their mother's chest, clinging to the long hair, and the older children rode on mother's back like a jockey.

[1] Baboons are usually born singly at any time of the year. The mothers show great affection for their babies, always fondling and cuddling them.

v. Young girl from Tanga in National costume, Tanzania.

One baboon remained behind to protect any stragglers from being snatched by a waiting leopard.

The crowned eagles[1] and the leopard are the baboons' most formidable natural enemy, and the mere sight of either send them into wild excitement as they utter deafening screams, barks and cries. (We had heard this in Tsavo and the noise was incredible.)

Often the leopard stalks a troop, waits for a baboon to stray and then pounces. Sometimes he has to leave the body and return for it later because baboons are extremely dangerous opponents, and six of them together are capable of tearing a leopard apart. (Instances have been recorded where an irritable domestic dog – such as an Alsatian – has launched an assault upon a baboon which has then clung to him with all four limbs and delivered horrible and fatal injuries in moments.)

In fact, as a leopard's teeth are hardly larger than those of a fully grown male baboon, he really prefers to attack while the troop is sleeping. Despite all-night sentinels the leopard generally makes such a kill quite safely, as the baboons are too afraid of the dark to take any defensive measures. Even here, predacity is essential, and where man has exterminated leopards from an area, the baboons have become a terrible menace to crops and vegetation.

During their normal day-to-day existence these powerful monkeys[2] congregate in large troops of twenty upwards. They live on berries, grubs, all kinds of fruit, leaves, birds' eggs, wild maize, spiders, insects and even scorpions, being agile enough to whip off the deadly tail before getting stung.

Their power of scent is little better than that of a human but they have extraordinary eyesight, and are among the very few creatures capable of discerning a carnivorous animal when it is standing motionless. Because of this faculty, other creatures

[1] The crowned eagle is more common throughout Uganda and the Congo. Both in appearance and hunting techniques this ferocious prey bird, with a wing span exceeding six feet, combines the most efficient features of the hawk, the owl and the other eagles. The pigmies regard this eagle as 'royal' and its beautiful feathers are held sacred; none but the tribal chief may wear them.

[2] Monkeys have tails. Those without tails (gorilla, chimpanzee, orang-outang, gibbon, etc.) are apes.

watch them carefully to see whether danger lurks around the water holes.

The baboons are regular visitors to the roof terrace at Treetops, and they can be fed safely as long as no attempt is made to touch them. Neither must a delicacy be withdrawn once it has been offered. Actually, the hostess prefers them to be ignored, otherwise familiarity breeds contempt on both sides. Guests are then liable to get bitten, through becoming less cautious; while some of the baboons grow increasingly confident and begin stealing food off the tables. For such rebels the hostess and staff keep a catapult and a small air pistol handy, the latter weapon having sufficient power to sting badly without inflicting any damage.

Of course, Jill was unable to resist encouraging their attentions, although she still has a scar on her arm from a monkey in Thailand which gave her a nasty bite while she was feeding it with a mango. The monkey had no intention of being malicious, but simply became over-excited. (Contrary to popular belief, the banana is not particularly favoured and monkeys adore juicy fruits such as mangoes, oranges, pears, etc.)

It was dark, less than thirty minutes after the monkeys had gone, and we all sat down at one long table to a gourmet dinner. The food is partly cooked at the Outspan Hotel and finished off in the small, spotless kitchen – where another catapult and air pistol hang – alongside the dining area. There was luscious asparagus imported from Italy, roast duckling and mounds of fresh strawberries with real cream. And for those who wished to watch *pink* floodlit buffaloes, there were unlimited supplies of free red wine.

However, eating was a hurried affair, because everybody wanted to get on to the verandah; strange noises and snortings were already emanating from the water hole. They marked a slight difference of opinion between a rhino and some buffaloes.

We had been close to rhinos for long periods on five occasions; but watching them uneasily, with one eye seeking out possible trees for a quick escape, is not the best situation for studying them. Desmond Vesey-Fitzgerald may have talked casually about dodging out of their way at the last moment, but frankly I would rather have a tree handy than rely on a nifty jump in the hope of avoiding two tons of armour-plated indignation.

It is said that rhinos have not a friend in the world, no sense of humour, no imagination, no benevolence, and are a perfect example of the fools who rush in where angels fear to tread. Eric Sherbrooke Walker calls them The Unloved Ones. Schopenhauer described their behaviour as irrational will-less perception. They have been termed Lost Souls, Wretched Wanderers and Clumsy Unhappy Clods. Recently, a modern philosopher even likened them to third rate boxers about to take the final count – Brawny, Brainless, Baffled.

Yet such malign descriptions always have an underlying sympathetic note; it is something akin to those who can see beauty in an ugly woman.

Certainly *I* found the rhinos fascinating. To me they were lovable louts. Both in behaviour and appearance they are creatures which run through an alphabetical gamut of gross contradictions. **A**wkward yet attractive. **B**lustering yet benign. **C**lueless yet charitable. **D**isgruntled yet dreamy. **E**ccentric yet endearing. **F**eeble yet fervent. **G**rotesque yet grand. **H**armful yet humane. **I**rritable yet inquisitive. **J**umbled yet just. **K**illers yet kindly. **L**ovable yet loathsome. **M**ighty yet muddly. **N**otable yet notionless. **O**bstinate yet orderly. **P**etulant yet peaceful. **Q**uerulous yet quiet. **R**evengeful yet reverent. **S**tupid yet sober. **T**urbulent yet tolerant. **U**ngainly yet unobjectionable. **V**aliant yet vague. **W**anton yet wary. **Z**ealous yet zestless.

Today, rhinos are among the rarest of animals. During the last 150 years they have disappeared almost completely from the face of the earth. In Java, Summatra and India rhino horn hunters have reduced their number to its current figure of around 600. (Throughout our months in the game sanctuaries of India we never saw *one*.) Mass slaughter since 1812 soon brought extinction in South Africa. There are a few left in the Sudan and the Congo, but in the entire world it is doubtful if there are more than 2,500 white rhinos left. East Africa, through the timely instigation of conservation programmes, now has the largest share, with their black bull rhinos reaching a shoulder height of 5 feet 8 inches and having front horns generally approaching 2 feet but occasionally up to 2½ feet.[1]

Before the last glacial period rhinos were abundant in five continents. The discovery of fossilized remains show they were often 13 or more feet in shoulder height, with horns of 6, 8 or even 10 feet. Just outside Ngorongoro, at *Oldupai Gorge*,[2] an immense variety of fossil fauna is being recovered from *Bed Number* 1 which, after scientific Potassium-argon tests, is considered to be 1¾ million years old. Already over 150 species of mammals have been identified and all of them were gigantic specimens; even the pigs had 5-feet long tusks and were the size of a hippopotamus.

Unlike most tusked animals, rhino horns are neither bone nor ivory; they are made of closely packed hair (similar to the tough fibre that surrounds a coconut kernal) growing from the skull. The front horn rises above the nostrils, while the rear horn almost covers the eyes; both are tough enough to inflict terrible wounds.

Now, the female rhino immediately below us was asserting her rights at the water hole. For a few minutes she would stand quietly, wallowing contentedly from side to side, in the deep mud. The resulting sideways movement was quite amusing; she jerked in rhythmically undulating ripples like a very old, very fat belly-dancer who, with a sad rheumatic jaunty manner, was indulging in a little nostalgic revelry, knowing that her days of male admiration were irrevocably finished.

Suddenly in the midst of her act she noticed the nearby buffaloes who, strangely enough, had politely turned their heads. Having poor eyesight she became confused by the black forms and annoyed at having an audience – she decided to investigate.

'What a damned nuisance, just as I was well and truly "squelched". To hell with them,' she thinks. Snorts, grunts and snuffles. 'Damn, they're still there.' More snorts. 'Oh, but this mud feels really great. It takes away all that dreadful skin

[1] In Amboseli there is a rhino known as Gladys who has front horns estimated at 4 feet.

[2] Here, in 1959, after twenty-eight years of personal excavation, Dr Jonathan Leakey discovered remains of early man and his culture. Since then, *Oldupai Gorge* has become one of the world's most famous and important archaeological sites.

irritation. Wonderful . . . but wasn't I going somewhere special? Now, where should I be going?' Silence. She squints. 'Damn, intruders. That was it. Must go and look.' Loud snorts. 'Perhaps they're only trees. This mud really is terrific. Funny, I haven't seen those trees before. Trees?' Grunts and sniffs. '*They're* certainly not trees. Damn, there must be others here. But who cares, I'm big enough to look after myself . . . I wonder who they are?'

From her actions it was almost impossible to hear the thought process of this lovable, blundering, grasshopper-minded colossus. (When being harassed by a rhino, their short memory and inability to concentrate is a great asset because the rhino often forgets where he is going and stops to graze before the charge is completed.)

Anyway, our female rhino took almost one hour to investigate the buffaloes, and every movement she made in their direction obviously entailed a tremendous effort. Withdrawing two tons from deep mud is no mean feat. Indeed, when aged or weakened, many rhinos have insufficient strength to climb out of the waterholes and remain firmly captured in the mud until they die and are eaten by the hyenas. This may sound an unpleasant death, yet just as wild life accepts predacity, the animals also appear conditioned to accept the inevitable end devised for them by nature.

The hostess told us that a few weeks before, an old rhino was bogged down at Treetops, and rather than watch the poor beast die they gently but firmly towed him out with two Land Rovers and a team of rangers. He emerged quite unharmed, but for some reason no longer showed any interest in life, moping around without eating and slowly getting weaker until the hyenas tore him apart while he was alive. His fate in the mud would have been quicker and kinder.

Our hostess now believes that when nature decides to foreclose on life it is best to leave the judgement unchallenged, and most conservationists agree with this policy.

Nature is wiser than man and if there is indeed an Almighty God then every bird and animal must have a purpose – perhaps equally as important as man – in the unfathomable complexities of His universe where even our own world can be no more than an epitome pattern of some infinite 'whole'. If such a pattern exists, then why should the animals not be closer to Almighty God than man? When death strikes from a source *within their own kingdom*, animals accept it graciously, yet man always fears death because he has no idea of what lies beyond. Perhaps the animals *do* know.

However, our particular lady rhino had plenty of energy and although having taken so long to investigate the buffaloes she eventually got within some yards of them. Puff. Grunt. Snort. She squinted again and then lowered her massive head until the front horn projected horizontally and she lumbered towards the nearest buffalo in a shambling trot. The buffalo also lowered its head (very slightly) and began inching forward as if to meet the threat. Even so, neither of them seemed truly perturbed.

Both were making a show of bravado. Both realized the tremendous power of each other. Both knew the horrible injuries a clash would produce.

Where such great power is balanced so equally most animals are prudent enough to settle any difference of opinion amicably, and our rhino halted long before the point of no return. Then, like a dispute between leaders of the Soviet Union and the United States, both animals huffed and puffed a great deal before turning and walking off in opposite directions; the rhino with its tail strangely looped in a question mark.

Even so arguments occasionally terminate in a real conflict; over the years, many visitors at Treetops have watched mighty battles when some elephant, buffalo or rhino has decided to push the 'red button'. Nearly always, the animals concerned have been mortally wounded. According to Eric Sherbrooke Walker, the original Treetops building used to shake violently when any of these three big powers[1] brought out their armoury in earnest.

Fortunately the elephant, buffalo and rhino have developed a high state of compromise and generally tolerate one another satisfactorily, apart from the inevitable bickerings and minor,

[1] Although a stray leopard very occasionally makes an appearance at Treetops, there are no large predators such as the lion. If the predators *were* encouraged, many of the other animals would become wary of visiting the water hole so freely.

uneventful squabbles which go on all night long around the water hole. But these squabbles are fascinating and the verandah at Treetops is ideal for detailed observation.

By 11.00 p.m. there were four rhinos and a calf, a complacent-looking creature that followed mummy everywhere. But once it got lost for a moment and, having the characteristic bad eye-sight of a rhino, it then mistook a buffalo for mum and tagged on behind. After a while mum looked back and, being very disturbed on seeing no child, began to wander around in wide circles with her head downcast as if searching for something microscopic.

In the meantime, the wandering buffalo had become aware of his unintentional adoption. He hurried and baby hurried. He changed direction and baby changed direction. He stopped and baby stopped. This was ridiculous!

The buffalo scrutinized the pool area and then most deliberately began walking towards the snorting, squinting mother rhino. Baby followed. The buffalo cleverly led the calf very close to mother and, after a cunning manoeuvre, baby was once again following its real parent, no doubt completely unaware that any change had taken place.

The buffalo's part in this operation was planned and intelligent. Nobody would be able to convince me otherwise, and it was just another thrilling example of how little we really know about the amazing capabilities of animals.

We had been lucky to see a calf because rhinos are slow breeders and it is believed they give birth only once in four years. With one-fifth of the rhino herds being poached yearly, this long reproduction period is yet another reason for their scarcity.

Also, rhinos exhibit a strange reluctance to move away from the locality where they were born. and this idiosyncrasy is an additional contribution to their depletion. Even when drought strikes, rather than change to another area, whole families of them prefer to starve and to die from thirst.

Likewise, the rhino is a city clerk type; year in and year out, his life follows a strict time routine that hardly varies. You can set your clocks by Mr Rhino. The same path, the same pasture, the same water hole and even the same selected place for depositing its dung. What an easy target the rhino makes for native poachers, who use arrows coated with a paralysing alkaloid distilled from certain plants, and similar in effect to curare, and then remove the horns with an axe, leaving the rhino to ooze a lymphatic bloody secretion, to fester and to rot. Its huge body, a miracle of nature, barbarically destroyed for a piece of matted hair.

Words and photographs are unable to describe the feelings inspired by the sight of such an atrocity. Unfortunately, such terrifying acts are too isolated from everyday life and to describe the horrible suffering of a strange creature slaughtered thousands of miles away, possibly raises brief public sympathy, but certainly no avid indignation and very little lasting interest.

But as long as rhino horn is thought to be a sexual stimulant, it is worth money, and the killing will continue. Poachers are always waiting and, if the rhinos continue to follow the same path to the same water hole at the same time, they can hardly hope for survival.

Like the Masai, these creatures cling stubbornly and tenaciously to a way of life which stems from an era when the plains and jungles belonged to nature. The habits of a rhino are from another age and make them far more vulnerable than most animals. East Africa may yet be the only place where they will survive.

We stayed out on the verandah until the sky was lightening, for even with dawn approaching the water hole remained busy. Although not a single elephant had appeared, the *Animals Seen* book[1] listed twenty-six for the previous night; and flicking through the pages their numbers occasionally ran into hundreds. This record has been kept for every day since Treetops received its first visitor on Sunday, 6th November 1932, and there are now a whole pile of these volumes; they include full details about unusual happenings, as well as the names of guests who were present at the time.

Naturally, there is no guarantee that any particular animal

[1] These records are maintained by experienced hunters who work at Treetops and keep an all-night vigil. The figures and details are therefore perfectly accurate.

species will appear at Treetops, but whether elephants, rhinos, giant forest hogs or hyenas, the sight is equally thrilling.

The entry for October 3, 1952 was particularly interesting because it marked the one thousandth party at Treetops. Guests from Tanzania, India, Sweden, South Africa, Tasmania and Kenya saw thirty-eight elephants, nine rhinos, nine buffaloes, seven giant forest hogs, six colobus monkeys and numerous plain animals.

I also noticed that on August 10, 1952, Prince Philip saw over 100 *baby* elephants (a record for calves) but no buffaloes!

Eventually we dragged ourselves away for a short rest before breakfast.

At 7.00 a.m. we were awakened by the hostess, and after everyone had been given coffee the Ford Transit arrived and took us back to the Outspan Hotel for breakfast.

Our sentiments about this visit were expressed most adequately by Peter Scott, when he wrote to Eric Sherbrooke Walker in February 1961: 'We shall certainly find ourselves talking about our night at Treetops very often indeed. It is a marvellous place conceived and created with tremendous imagination.'

CHAPTER XVIII

Bill's Place

Opposites – Perfect balance – A fantasia from Hollywood – For the ornithologist – Nights to remember – Another time dimension – Life has a meaning – William Holden – Sportsman superb – The days pass – Dominant glaciers – Experienced mountaineers only – Dinner by candlelight – A thrill for the ladies – No comparison – In a nutshell.

THE Mount Kenya Safari Club is quite close to Treetops. Although they have equal charm, no two places could be more unlike.

Accommodation at the Club is provided by one main building plus individual 'chalets' set among acres of artistically landscaped grounds.

The scene does not have the uncultivated naturalness of a National Park yet there is nothing to suggest the uninspiring neatness of public gardens. A perfect balance has been achieved between man-made beauty and wild Africa.

Sited in a valley almost at the foot of Mount Kenya, the Club's lakes, the billiard-table lawns, the undulating wooded slopes and the casual masses of nurtured tropical flowers all harmonize with luxuriantly vegetated countryside. In the background the twin snowy peaks of *Batian* and *Nelion* rise to over 17,000 feet. This fantasia looks like something straight from a Hollywood film epic, but then the Club's President *is* William Holden.

There are no buffaloes or elephants wandering around; but an extremely wide variety of resident bird life will delight even the experienced ornithologist.

The Club is not ruined through being pompous, formal or

'snobby', although everything about it is luxurious and expensive, including the charges! One is never aware of being overwhelmed by luxury. Never aware of ostentation. Only the overall effect is predominant. Small touches like rifle-shaped lamp holders and an impressive, simulated tent décor for the dining room, add the final touches of safari life.

Accommodation in the main building is gracious, with every bedroom having ample lounge space, comfortable chairs, attractive furniture and the atmosphere of a personal African home.

The nights are chilly, and on retiring for bed we were delighted to find that a log fire had been lit in the enormous stone grate. With the room enveloped by its cosy, warm glow, we relaxed silently, just watching the sparks crackle up the chimney. Occasionally, the logs emitted a stream of combustible gas which would then ignite and spit out a hissing, pale blue flame that filled the air with sweet fragrance.

Outside, beneath a full moon, the lower slopes of Mount Kenya were silhouetted in dark purple. Much higher, the wooded landscape changed to a silver yellow, and higher still, it became mauve until the escarpment was hidden finally by an impenetrable blackness.

Moments such as these bring an indescribable tranquillity and the mind skirts another time dimension where some missing cipher is keyed into the whole world of incalculable numbers, and with a transitory thought pulse, the mystery of life has a meaning.

Sadly, William Holden is a hunter, but at least his safaris are long and tough, often entailing exploration deep into the bush. There are no air conditioned tents and no professional trackers or native carriers, for these safaris are a personal challenge between William Holden and the big game he is out to shoot. Certainly his method of hunting complies with the true definition of a sportsman and notwithstanding the use of a firearm he most certainly *undertakes a daring pursuit and will give his opponents a fair chance.*

Even a conservationist can say 'good luck' to any man whose hunting incurs real danger and hardship as well as requiring considerable skill. Under these conditions, the outcome of such a challenge to nature is not a foregone conclusion, but unfortunately few hunters are of the William Holden calibre.

The days at the Club pass leisurely. The air is cool and invigorating, the sky is azure, and the landscape colours are vivid. In this mountain retreat there is nothing to do except relax and unwind jangled nerves. We walked amidst the cranes, peacocks, bearded woodpeckers, larks, wagtails, parrots, turacos, hornbills, warblers, swallows, honey guides and hundreds of other species which, although absolutely free, are so tame they will take food from the hand. We sat and watched the ever-changing hues on the glaciers of Mount Kenya[1] where harsh, dominant rock faces twist upwards like a bizarre iced cake.

At night, log fires burn in all the rooms. There is an intimate bar and lounge where social barriers do not exist. Dinner is by candlelight. And when William Holden is in residence, as he was when we were there, he mixes with his guests just as Dr von Nagy had done. For the ladies, this in itself appears to be a thrill quite equal to the grandeur of Africa; it was for Jill!

This, in a nutshell, is the Mount Kenya Safari Club. It is not like a Stanley Hilton or the Club on Mafia Isle or any of the Game Lodges. And the scenery is entirely different from Meru, Ngorongoro or Tsavo. Everything about this unusual retreat is quite incomparable. It is simply – *Bill's Place*.

[1] Author's Note: Mount Kenya can be climbed only by very experienced mountaineers. It is a tough, extremely dangerous ascent. The easiest route – if it can be called such – is by the southeast face and involves crossing a deep glacier followed by five hours of precipitous rocky cliffs to *Nelion* (17,022 feet) from where the highest summit of *Batian* (17,058 feet) is reached after a tricky passage through the *Gate of Mists*. Even to reach the snow level at about 14,000 feet takes two to three days. However, on the lower slopes, below the northern face, an eight-hour circular hike takes in the Mackinder Valley, and this is said to offer the most spectacular mountain scenery in the world. Camping sites are available here.

CHAPTER XIX

Interlude

A sad parting – 30,000 feet above Kenya – Technical superiority – No faith – Comfort and reliability – Unrivalled ingenuity – BUA's impressive history – Unfortunate jokes with an eccentric – Disgrace – Icebergs and sugar puff cakes – Entebbe – That extra service – Three welcomes.

FROM the Mount Kenya Safari Club we returned to Nairobi where we parted with our Cortina, for some of the roads in northwest Kenya were already badly flooded by premature rains. With the dry season fast ending, we decided to save time by flying the 928 miles to Entebbe in Uganda; from here we could hire another car.

Even so, it was sad to leave Goliath. She had become a vital part of the team, a personality; a little heroine from Ford who, without a murmur of resentment had suffered terrible punishment for thousands of miles on the worst roads imaginable, where we had seen numerous foreign cars grind to a mechanical halt.

Now I was once again feeling national pride for we were 30,000 feet above Kenya in a VC 10 operated by British United Airways. A magnificent product of the British Aircraft Corporation, our VC 10 is the finest aeroplane in the sky today. Four ultra powerful Rolls Royce Conway engines are combined with unique wing flaps to give an unrivalled performance: short field take-offs with a cruising maximum of 600 m.p.h. and a landing speed no faster than a Viscount, while revolutionary construction features also make it the safest aeroplane flying. All British radar and navigational systems – acknowledged as the most advanced available – complete its technical superiority.

Strangely, BOAC believed the range and payload factors of the VC 10 were unsuitable for profitable operation, and ordered Boeings. Many foreign airlines followed BOAC's example, and how wrong they were! Of course, BOAC now flies the VC 10, and even wish they had more in operation, but export figures have never achieved the success that such an aeroplane deserved.

On the other hand, British United Airways ordered VC 10s without hesitation, and as this airline also offers a service that is second to none for comfort, reliability and technical know-how, I was more than happy to be aboard this particular aircraft.

Britain could do with a few more people and industries showing national pride; it is not an unworthy vice, and certainly no country warrants greater pride, for British technology and ingenuity is unequalled. Let nobody doubt this fact.

BUA have an impressive aviation history dating back to 1928, when they built and operated one of London's original airports. Now their routes cover many resorts in Europe, the Canary Isles, East, West and Central Africa, South America and also from Gatwick to Scotland. I have flown with them on many occasions and find them the best of our British airlines, but then they are *truly* British in every sense of the word.

Thinking about my original flight from London to Nairobi I could hardly restrain my laughter. I had sat next to a terribly-terribly British gentleman who, already dressed in a green safari outfit, was obviously a rich eccentric. During conversation he asked where I was staying in Nairobi.

'Ah, the New Stanley,' he said, nodding with approval and speaking in a staccato, telegraphic manner. 'A wonderful hotel. Wonderful. Very central. Safaris coming and going all day long. An ideal place. The Lions meet there, you know.'

'Really?' I must have looked *and* sounded incredulous.

'And the Round Table and the Rotary Club,' he added tactfully, but giving me a fierce look which showed he realized my stupid misinterpretation of the Lions.

After that he ignored me for quite a time, but at length he spoke again. 'What Lodges are you visiting?' The question was fired, not asked; fired like an unexpected rifle shot.

For a second I was stunned. *Thinks: Musn't make a fool of myself this time. Lodges? What the hell is he talking about now?*

'Do you mean Masonic Lodges?' I answered after an unnaturally long pause.

He went crimson. 'Masonic – Masonic – MASONIC.' Each time the word was repeated it was louder and with an ever increasing hiss. 'I meant *Game* Lodges. Game,' he spluttered.

We never spoke again. Obviously Africa was not yet in my blood, and for the remaining part of the journey I was left in idle peace to contemplate my 'disgrace'.

I watched the cloud formations which, on our landing descent, first appeared as solid grey mountain ranges, then broke into glistening icebergs scattered over a blue sea and finally became no more than isolated sugar puff cakes that were whiter than white.

The airport at Entebbe is small. A BUA hostess helped us with our hand baggage (another hostess carried someone's baby – touches of that 'extra' service) which, with Eumig, Rolleiflex and Canon cameras, plus a typewriter and several large wooden carvings bought by Jill in Kenya, was rather an excessive load.

A direction indicator told us Timbuktu was 2,654 miles aways London 4,026 miles and New York 7,065 miles. Twin signs, read *Welcome to Uganda* and customs officials extended the welcome in practical terms with broad, white-toothed smiles.

This made several distinct welcomes – Signs, Smiles and Sun.

CHAPTER XX

Uganda – Cradle of the Nile

A fairytale – An ancient quest – Speke discovers Lake Victoria – Historic diary notes – The Nile jigsaw puzzle solved – Thunder and foam – Mobile again – A duel – Acrid fumes – Fifty miles from nowhere – Hippos – Sly enemies – Tender crocodiles – Backwaters – Prolific bird life – Disease – Swept downstream – Almost capsized –A gigantic pressure hose – Beauty can be ruthless – Need for caution – Nature at her magnificent best – The tranquility of life – The power of death.

ALMOST one-eighth of Uganda's 94,000 square miles is open water, which comprises mainly Lake Victoria from where the River Nile begins its 4,160 mile journey to the Mediterranean. To the north of Uganda lies the Sudan and to the west the the Congo. Tanzania and Rwanda form a southern border, while the eastern sector adjoins Kenya.

'The country is a fairytale. You climb up a railway instead of a beanstalk and at the top there is a wonderful, new world. The scenery is different, the vegetation is different, the climate is different and the people are different; in fact, it is all different from anything else to be seen in the whole range of Africa . . .,' wrote Sir Winston Churchill in *My African Journey*.

Entebbe, a city of gardens and parks, is situated at an altitude of 3,863 feet on the shores of Lake Victoria, which is three and a half times bigger than Wales and the world's second largest lake. This is the 'Cradle of the Nile' and from ancient times 'Caput Nile quaerere' (to discover the source of the Nile) became the quest of many adventurers.

Indeed, the desire to find the Nile's source dates back to the fifth century BC when Herodotus, a Greek historian, took a great interest in why the Nile rose every summer for about 100

VI. Murchison Falls where the Nile thunders between a rocky gorge, Uganda.

days and then subsided again. The 'learned' people in Egypt failed to offer Herodotus any possible explanation, and no progress was made until the late nineteenth century when John Hanning Speke and Richard Francis Burton discovered a huge inland sea which was subsequently named Lake Victoria.

Being convinced this sea was the Nile's source, Speke organized another expedition and returned with James Grant to prove his theory; on this occasion they discovered Lake Kagera. Then Speke and Grant took different routes, and after several days Speke found the Ripon Falls pouring into a fast flowing river, the site of which is now marked by a stone monument, bearing the inscription *Speke Discovered This Source Of The Nile On The 28th July* 1862.

Speke was so excited by his discovery that, instead of making the usual brief diary entry, he recorded the scene with much greater verve: 'It was a sight that attracted one to it for hours – the roar of the waters, the thousands of passenger-fish, leaping at the falls with all their might; the Wasoga and Waganda fishermen coming out in boats and taking post on all the rocks with rod and hook, hippopotami and crocodiles lying asleep at the water, the ferry at work above the falls, and cattle driven down to drink at the margin of the lake – made in all, with the pretty nature of the country – small hills, grassy topped, with trees in the folds, and gardens on the lower slopes – as interesting a picture as one could wish to see.' (The ferry mentioned by Speke had been in operation for a long time linking Busoga with Buganda.)

But Speke's discovery was no more than a 'beginning'; just a small link in a very big chain. Only nine months later, Sir Samuel Baker located a *second* source (Lake Albert) and during 1874, Charles Chaillee-Long discovered a *third* source in Lake Kyoga. Then, while H. M. Stanley was leading the first European expedition to Uganda's 17,000-feet high Mountains of the Moon, he also explored the upper waters of Lake Kagera, and established that Lake Albert was connected by a river with yet another lake.

For years, independent geographers were continually finding new pieces which belonged to this complex Nile jigsaw puzzle that had baffled and defeated explorers for centuries; not until around 1900 were all these pieces eventually fitted together.

Now, the mosaic of rivers, lakes and cataracts form a known pattern: The Kagera River and its tributaries descend from the 12,000-feet high Virunga Mountains[1] and enter Lake Victoria at a height of 3,400 feet. Near Jinja, the foaming Nile[2] surges from Lake Victoria, flows through Lake Kioga and then, 100 miles north of Jinja, after being compressed by nature through a 19-feet wide rocky gap, the waters thunder onto the rocks 140 feet below, producing the famous cataract known as *Murchison Falls.*

After Murchison the Nile is comparatively peaceful until reaching the Sudan border, when once again it becomes a mighty cascading river. It swirls through the Sudan in a series of violent rapids, and thereafter has numerous mood changes on its long journey.

Within one hour of arriving at Entebbe airport we were on our way north (in another white Ford Cortina) making for Paraa Lodge on the edge of the Nile.

After Goliath's admirable performance I had resisted the recommendation of a self-drive Mercedes and insisted upon another Ford which I was delighted to find fitted with Firestone tyres again (except the spares) for these had also proved their worth for our thousands of miles through Tanzania and Kenya.

Really, I suppose their performance was hardly surprising because Firestone have pioneered the tyre market for sixty-five years and their present organization employs 86,000 men and women in twenty-three countries. During 1906 Ford Motor Company gave Firestone the largest single order ever placed by the motor industry and since then, Firestone's progressive history has been highlighted by more than a dozen major technological 'breakthroughs' that have revolutionized the whole industry and include developing high speed aircraft tyres for World War II. Now, the VC 10 jet age has brought

[1] The Virunga mountain range extends sixty miles into Uganda from Rwanda.

[2] From Jinja, the Nile is known as the *Victoria Nile.* After leaving the northern end of Lake Albert it becomes the *Albert Nile.* From the Sudan border the name changes to the *Bahr El Djebel* and just before Sudan's Lake No, the famous *White Nile* begins.

even greater problems, but in 1952, Firestone produced the very first tyre that could withstand constant landings at 250 m.p.h. and this met every requirement establishment by the Air Force.

For an organization having such impressive technical capabilities and with a network of laboratories and test centres all over the world I should imagine that Firestone found little difficulty in manufacturing tyres suitable to tackle the worst roads Africa could offer. Ford and Firestone. They had certainly made a reliable team for *our* safari.

After twenty-one miles we came to Kampala, Uganda's capital, but as this could be explored much later we drove straight through, continuing towards Nakasongola (about one-third of the way to Paraa) on a road surface that alternated between something like mud wallows and a ploughed field.

In a small village – the kind which you miss while batting an eye – we stopped at its only store to buy some food and Coca-Cola for a hasty picnic meal. Then I realized the car hire deposit had taken all our remaining currency and we possessed only travellers' cheques.

The store was managed by an Indian and, being typically shrewd, the travellers' cheque I offered was studied closely, even held up to the light.

'I am sorry, very, very sorry, but this is not acceptable for payment,' he informed me after the detailed examination. 'If you have cheques issued by Mr Thomas Cook or the American Express I will take those very very gladly.'

'But this is a Barclays cheque and it's just the same as any other.'

'Are you Mr Barclay?' he asked.

'No, this is Barclays Bank of England. Surely you know the name. There are Barclays Banks all over Africa.'

He shrugged his shoulders disbelievingly.

'There are even Barclays Banks in Calcutta,' I added hopefully.

'Very, very sorry,' he repeated. 'Mr Cook or American Express only.' (We had experienced an almost identical situation to this while in India, even to being asked if I was Mr Barclay!)

Then, just on the point of resigning myself to thirst and hunger, a faded blue notice caught my attention: *Diners' Club. Associated with the Standard Bank Limited.*[1] Surely this tiny store, miles from anywhere and managed by an Indian who had never heard of Barclays Bank, could hardly be a member of Diners' Club? Yet it was!

Obviously the Indian had never handled a Diners' Club card before and for some reason now regarded me with great reverence. However, there was some delay in finding the special franking gadget, a further delay sorting out the necessary carbon form, and a final delay while I showed him how to use the 'stamping' gadget. Nevertheless we were getting food.

It was not the first occasion that being a member of Diners' Club had helped us out of a difficult situation. Holding such a credit card is like owning the proverbial million-pound note; you need only sign bills to procure goods and services all over the world – airline tickets, petrol, television sets or anything else. Even a loaf of home baked bread, six bottles of Coke, a tin of luncheon meat, some overripe fruit and two bars of partly melted chocolate, which were the impressive delicatessen commodities my signature had just purchased for our snack.

Just outside the village we stopped to eat our feast. As daylight was fast running out, we had had no intention of stopping, but sudden torrential rain made driving visibility impossible.

For several hours the air had been getting hotter and deoxygenated. Then we breasted a hill to face black gloom sweeping towards us from the east and, in the tropics, nothing produces more vivid contrasts than a duel between sun and storm.

On one side of the road, the flowers enjoyed an unusual brilliance and the buzz of crickets swelled to a crescendo. But

[1] Author's Note: The British Diners' Club operates from 214 Oxford Street, London, W.1. Membership is unrestricted and costs three guineas a year with no other charges involved. Diners' Club are responsible for paying all bills signed by their members and subsequently these members are then reinvoiced. I have no special connection with Diners' Club, but having been a member since 1959 I find it operates an excellent credit system which has a far greater coverage than most competitive organizations. Unfortunately many people overlook this valuable service because they associate it only with wining and dining.

on the opposite side all colour was replaced by a sombre grey and the insects barely hummed. A reptile shot across our path to the sunny west, changing from an unrecognizable leaden body into a flashing blue-green lizard.

The storm broke with violent rain which hit us like a wave. Fork lightning tore the clouds apart and a deafening thunder crash shook the car. One charge stabbed to earth in a crackling spear just yards away; it filled the air with acrid fumes and woodsmoke.

After thirty minutes of constant fork lightning and another couple of high voltage spears, the short circuit was isolated, the contact breakers were withdrawn and the air cleared. Also, the blown fuses had been repaired and once again sun shone upon the landscape which was now steaming and alive with tiny frogs splashing in the rain pools.

Paraa Lodge is fifty miles from any civilization and overlooks the Nile; throughout the night it was thrilling to hear the 'haw-haw-haw' bellows of hippopotami. Early the next morning we were chugging close to them in a small wooden boat.

A hippopotamus is the biblical behemoth,[1] and at one time these enormous creatures lived on every river bank and lake shore in Africa. But now, even along the Nile from the Delta to Khartoum, these good-natured giants have been foolishly exterminated, shot from pleasure steamers by those tourists who as yet fail to appreciate the more subtle thrill of shooting with a camera.

For the natives, this extermination has sometimes been disastrous because hippopotamus dung stimulates aquatic life and thereby produces an abundance of fish. Many Nilotic tribes are entirely dependent upon the River for their food and when the placid hippos disappear, so do the fish.

At present around 30,000 hippos are under protection in Uganda and certainly there was no scarcity of them along this particular stretch of the Nile, although they are still poached extensively for their tusks which render a mere 6 lb. of ivory.[2] Less than ten years ago, hippo's skin was also in great demand

[1] The word is assimilated to the Hebrew plural – *of dignity*. (See Job XL.15.)
[2] Hippo ivory is much softer than that of elephant.

for a formidable type of whip known as the *sjambok* in South Africa and the *kiboko* in East Africa.

During the day, hippos spend most of their time browsing in the water but when completely submerged they must surface for air about every three and a half minutes. Then a gigantic pink head emerges and water is blown powerfully from their nostrils followed by a rumbling bellow. Swoosh! 'Haw-haw-haw'; the noise was familiar.

As hippos prefer to sleep below the surface, a sensory organ automatically brings them up for air at the required time intervals and having taken oxygen they reimmerse with equal automation, remaining asleep throughout the complete process.

While basking the nostrils are kept just above the water, and apart from beady eyes and ridiculously tiny ears, nothing else is visible. Certainly the eyes and ears give no idea of the enormous head to which they belong; and likewise, the head is not so enormous that it indicates the colossal size of a hippos body which may be 14 feet long and weigh 5 tons.

To watch a hippo leave the water entirely is quite an experience when seen for the very first time. The river is empty – calm – tranquil. Then something disturbs the surface. A fish perhaps? Slowly, a pair of small eyes and even smaller ears appear. 'Ah, it must be a type of otter', you think.

The water swirls and foams. A tremendous head emerges, square and broad with nostrils seemingly large enough for posting letters in. Swoosh! As if 'blowing off' a massive snorkel tube, water jets into the air under pressure, leaving a blanket of spray falling like heavy dew; momentarily a rainbow forms. 'Haw-haw-haw'. (The *Queen Mary*'s siren – which could be heard over a distance of eight miles – was no more overpowering than this foghorn bellow.)

After the 'hawing' comes a yawn. The mouth opens – opens – opens and continues to open; there is a little more to this elasticity than *two-way stretch*. Such a black cave inspires thoughts about Jonah, about man-eating clams and bizarre pillar boxes that swallow mini cars. At least the *Ring of Confidence* is there, with four curved ivory 'teeth' protruding from the front corners of each jaw; the largest incisor is probably about $2\frac{1}{2}$ feet long although a record hippo tusk has reached 3 feet $4\frac{1}{2}$ inches.

Even at this stage, the hippo's true size is only appreciated when he climbs on shore and a fleshy pink body, somewhat larger and heavier than a car, pulls itself from the water to move across the land with surprising agility on stubby legs.

We know of this agility because a hippo unfortunately surfaced while partly beneath our boat. The stern lifted from the water, and then we dropped back with a crash. For a moment I thought the keel had caught some rocks and we were about to capsize.

This slight collision really scared the poor hippo. He raced to the shore as if jet propelled, giving a fine example of the superb swimming capabilities which enable them to cross the straits between East Africa and the island of Zanzibar. Once out of the water he galloped off through the bush, leaving a trail of dust. The hippo's ancient Greek name which meant *Water Horse* was obviously an apt description.

Apart from encouraging fish life, hippos perform another vital service by keeping in check the aquatic vegetation of papyrus and sudd, which would not only severely impede navigation, but also cause floods and a whole network of undesirable channels. Then the clogged waters encourage the breeding of malaria-carrying mosquitoes and other equally unpleasant insects. This is comparable with the service rendered by elephants who maintain a status quo in jungle vegetation, and prevent many land areas becoming inaccessible.

Where the hippos have been destroyed, many stretches of the Nile already suffer badly from such problems.

In effect, the hippo can be regarded as a giant dredger, and walking along the bed of a river or lake he maintains orderly pastures which are skilfully harvested to provide his daily food ration of around four cwt. Babies take some months to learn this technique of 'river bed harvesting', and at first mother leaves them concealed among the bankside undergrowth. Later they learn to suckle while under water, but laboriously have to surface for oxygen between every mouthful.

Although such good-natured, inoffensive beasts, two hippopotamus bulls will often fight to gain supremacy over a certain territory. The dispute generally begins with the animals showering each other with excrement, shooting it directly from the anus and using their tail to spread a foul smelling cloud of it over their rival.

Following this unusual 'smoke screen' ritual, their powerful tusks are then brought into a battle which may last for several hours. These tusks are used only against their own kind; hippopotami have no enemies; their skin is impenetrable even by crocodiles or lions, yet the insignificant water leaches do cause them a great deal of unpleasant irritation.

At night, hippos occasionally will attack a camp fire (if it is burning close to the river) or the lights of a vehicle, while some very old cantankerous bulls *have* been known to start deliberately upsetting river craft, but usually they exhibit friendly, peaceful dispositions . . . Swoosh! 'Haw-haw-haw.'

I hoped there was no such cantankerous bull in this area because our decrepit outboard motor had now broken down and there were too many tree trunks floating around for my liking. These were the dorsal vertebrae of crocodiles; 10 feet, 15 feet or more long and weighing around 400 lb. Great predatory creatures which descend from a prehistoric species that for millions of years ruled the rivers and lakes.

Strangely, early man felt no animosity towards crocodiles because he apparently endowed them with a dignified aura. This is indicated by archaeologists finding these reptiles mummified in Egypt, and also by explorers discovering crocodile 'god' images throughout parts of Africa. Even today, there are many tribes – some in East Africa – who still endeavour to 'charm' their own particular crocodiles and to lure them ashore by hypnotic chants and songs, just as the Namuana villagers enticed the two giant turtles *Raudalice* and *Tinaicaboga* on the remote Fijian island of Kadavu.[1] And it is claimed that, by whistling a specific tune on a reed, certain Ugandan natives were able to attract a 25-foot long crocodile onto the banks of Lake Victoria, only fifteen years ago, where the Entebbe Golf Course is now established.

Hungry crocodiles are sly, cunning enemies and no matter

[1] Described in *Tropical Quest*.

how cautiously animals come to the water's edge for drinking, there is always the time when an almost wholly submerged croc remains undetected.

A gazelle approaches slowly and timidly. The invisible crocodile watches and waits, his larynx and oesophagus closed by a valvular skin seal and his bulging periscope type eyes at surface level. A fast, deadly, reptilian submarine with its *tail swipe* missile already sighted on the target.

The gazelle stoops to drink. Quickly and silently the crocodile manoeuvres for an attack: 'Full speed ahead. Maintain periscope depth. No surface noise. Fire one . . . fire two.' Fearful jaws close on the creature's head and a powerful tail swipe knocks the gazelle right into the water. Apart from some bubbles and a little froth the mission is accomplished.

As crocodiles cannot tackle large prey until the flesh putrifies, a victim's body must be dragged to land where it is hidden and left for several days while decomposition takes place. During this time the crocodile lives on a meagre diet of tiny fish.

Had large fish been available in the first place the gazelle would be still alive, for the croc prefers to catch a nice fat perch and eat it immediately than wait for weeks for a meal of decomposing animal flesh. But where the hippos are exterminated and perch is scarce the crocodile has been forced to change its feeding habits.

(Indeed, wherever man disturbes the balance of nature, this kind of thing happens, and animals have been known to alter their habits entirely. The West Indian mongoose is a perfect example. Many years ago, an island in the Caribbean imported hundreds of additional mongooses to help destroy their snakes. Nature's balance was thereby upset; snakes became so rare that the mongooses turned into chicken plunderers. Then the mongooses were exterminated by mass poisoning but the island was besieged by frogs, mice, rats and other rodents which the mongooses had hunted. Eventually, to prevent catastrophe, more mongooses *and* snakes were imported, but it was a long time before the balance was restored.)

Once crocodiles have eaten, they become sluggish, lazy and peaceful, basking in the sunshine and letting Egyptian plover birds peck insects from between their scales. The plovers also have a reputation for picking remnants of food from the crocodiles' teeth but we never saw this happen even once and believe it to be a fallacy.

The crocodile is hardly impartial to investigating a nice human leg simply out of curiosity and, as our efforts to start the outboard motor were rocking us about considerably, I was very relieved when the exhaust began pumping out a continual stream of black smoke. Once more we set off, despite the mechanically unhealthy fumes and a tendency to misfire when the throttle was more than half opened.

As we passed island mud banks, disturbed crocodiles slid into the water as smoothly as snakes, keeping their bodies low to the ground and disappearing below the surface without a ripple. Others remained absolutely still with their great mouths wide open, their gullets pink, their armour plated scales blending with the mud and their yellowish teeth like the spikes in mountaineering boots. They were reptiles turned into cold stone by a Medusa. Sinister carvings. The sight of them made me shiver.

But even crocodiles are not all malevolence and they have a tender side to their nature. The female fans out her forty to sixty eggs in the shape of a necklace, and after burying them beneath sand or rotting leaves she guards them with great diligence. The incubation period is around ninety days (from December to March). When the babies are about to hatch they make high-pitched squeaking calls which attract mother's attention so that she can help to break the shell and clear away the surrounding mud. These squeaks have a rising and falling inflexion which follow a distinctive wave-form pattern, and this is the rhythmic noise that native 'charmers' try to capture in their music.

Once out of their shells, few of the babies survive in spite of a helpful, protective mother. On land, mongooses and monitors[1] are waiting for these newly hatched delicacies. From the sky,

[1] Dark green, yellow-striped reptiles that reach a length of 6 feet. They can swim, dive, move quickly over land and even climb trees. Crocodile and tortoise eggs are their favourite food, and the female monitor gains so much experience during these 'egg robbing' activities, that she protects her own eggs by laying them in a termite's nest where they are comparatively safe. But the monitor lizard is also practically extinct, being killed for its beautiful skin which has trophy value.

marabous, whale-headed storks and other predator birds swoop on them greedily. And those babies who evade the attacks often reach the water only to be gobbled up by a fish.

The survivors grow approximately 1 foot in length every year and do not become sexually mature until they are about 8 feet. After ten years, their rate of growth declines but never stops completely and an eighty-year-old crocodile would reach somewhere around 25 to 30 feet long.

Nowadays, the Murchison Falls area is the only place in East Africa where crocodiles are plentiful. Most of their colonies were exterminated by uncontrolled hunting before any of the conservation areas were defined.

Crocodile skins are a lucrative market and prior to 1950 hunters travelling through Uganda had already eradicated them from Lake Kioga and Lake Victoria at the rate of 13,000 a year. They are quite extinct in South Africa, virtually non-existent in Egypt, and another five years will see them gone from Rwanda, the Congo and the Sudan. The skins are used for making shoes and handbags, but only one 15-inch portion of the belly is utilized and it must be culled from crocodiles between 5 feet and 6 feet long. This portion is worth about £15, which to native poachers is a great deal of money.

The idea of 'farming' crocodiles has been explored by several businessmen but the greedy crocodile is an expensive monster to rear. It takes 12 lb. of protein to build 1 lb. of croc, who devours about 1,500 lb. of food before maturing enough for his skin to be used by the leather industry. With the high cost of fish such ventures are uneconomic and so poaching continues at an ever increasing rate; Paraa Game Wardens use a Triumph land/water car for patrol checks.

Crocodiles may appear sinister reptiles but nature herself keeps their numbers under very strict control and if man has inadvertently turned them into flesh-eating monsters, the poor old croc can hardly take all the blame.

To wipe out this prehistoric reptile species would be a biological tragedy especially as the crocodiles at Murchison Falls indirectly help to conserve other animals. Because they are such a big attraction, every tourist who visits Murchison puts more funds into the national parks. As Alan Moorhead wrote: 'Even the crocodile, it seems, is not the outcast of nature that one had formerly supposed it to be.'

The misunderstood crocodile therefore not only has a monetary value to conservation programmes and an unexpected tender streak, but the stories of crocodiles shedding tears over their victims are quite true. All amphibious saurians experience a fast evaporation of the blood's salt content, and this causes fluid to be eliminated through their eye glands. Using energy speeds up the evaporation process, and after making a 'kill' the effect often produces an absolute flood of excess fluid – real crocodile tears!

We continued on our journey down the Nile, enjoying an infinite variety of scenery: along shaded backwaters where the reed masses hardly moved in a lazy flow; close to banks of tropical flowers and lush vegetation; past fierce jungles where tamarind, acacia, trichilia, balanites and ambatch[1] trees leaned out over the water, often causing us to bend our heads; into fast-moving currents that frothed between island banks of mud or sand where stunted palms grew; sweeping down wide stretches bordered by light rain forest; ploughing through channels almost blocked by papyrus and emerald green Nile cabbages floating on the surface together with white, purple and crimson water lilies.

Mating dragon flies of a metallic lustre danced a *pas de deux* just above the surface. Butterflies, absorbing moisture from the river banks, rose in a blue snowstorm.

A dazzling white-headed fish eagle zoomed, rolled and looped through the air in a thrilling display of acrobatics. These birds cannot fly if their feathers become wet and consequently they can only take smaller fish, but often a greedy eagle can be seen struggling towards land with quite a large perch. Even if the load is really too heavy, the bird is usually unwilling to release his catch and would rather fall and drown than forgo the anticipated meal.

A rare black crake crossed a narrow channel of water by

[1] Ambatch trees, together with papyrus, have been known to block the Nile so effectively that their combined 'debris' has made a bridge strong enough for elephants to cross.

striding from one Nile cabbage to the next, leaving us with only a fleeting impression of long, fiery red legs and a vivid yellow beak. Ash-brown Egyptian geese circled the papyrus in a tight formation before spreading their wings and dropping to the ground.

Now two white pelicans were skimming the water on a fish reconnaissance; almost at stalling speed they lumbered along as if having engine trouble and looking for a place to land. In contrast, they take off from the water in fine style. Huge but graceful birds, silverish wings move faster and faster, churning a wake of white foam, then suddenly they lift themselves free and climb in a steep turn. They can also land just as dramatically, gliding down at speed, dropping their legs on to the surface with the expertise of a champion water skier.

Storks, herons and marabous were standing on the mud flats waiting erect and motionless. From the undergrowth a blue face with an ugly, outsized bill scowled at us and quickly disappeared; I think it was the shoebill, a retiring bird very rarely seen.

Then the noise of our motor frightened a secretary bird. It gets its unusual name through its resemblance to a Dickensian clerk wearing black trousers, a white shirt and a quill pen behind both ears. (When it gets excited the quills erect to form a beautiful crest.) Usually found singly, they strut along as if on stilts, capturing grasshoppers, snakes or lizards by pinning them beneath a foot which shoots out with unbelievable swiftness. Because these birds are nearly always solitary, natives[1] call them *labang-cula*, which means *He Who Wanders Alone In The Wilderness*.

Although a powerful flyer, the secretary bird needs a prolonged take-off distance, and it was amusing to watch him running clumsily with his wings outspread. He finally gained enough speed to get airborne, and left the grass like an overloaded bomber; but once off the ground he climbed away quite easily.

For biologists and tourists, Murchison is a paradise, but those riverside peasants who frequent the Nile area in Central

[1] Author's Note: In Uganda the word *native* is disliked and always associated with primitiveness. The preferred term is *national*.

Africa have a particularly unhappy existence. First, malaria is endemic, and most of the natives are permanent victims of this disease. They experience a recurring fever which produces alternating phases of violent shivering and furious sweating. Every limb feels a leaden weight and excruciating pains shoot through both eyeballs like a knife being twisted. The head throbs. Where no medicine is available, the native eventually dies.

Sleeping sickness and hookworm also take a heavy toll. Then there are jigger fleas, which lay eggs beneath the skin and cause terrible itching, and the *mbwa* fly, a tiny black insect that introduces weakening parasites into the blood stream which can cause blindness.

But by far the worst ailment is bilharzia;[1] though at least it is an infection that can be avoided because the parasites responsible live only in fresh water, and people must obligingly offer themselves to the scourge. Ponds, streams, even Lake Victoria, and practically every river from Cape Town to Cairo is infected, so any bathing or swimming is to be avoided. Even a ten-second paddle will give the parasites ample opportunity to penetrate the skin.

The disease has an ancient history, and is not confined to Africa. The parasite existed over 6,000 years ago – proved by the medical examination of Egyptian mummies belonging to various early dynasties. Nowadays it is also prevalent in Arabia, Iraq (the Tigris and Euphrates are heavily contaminated), Egypt, the West Indies, parts of South America, China and Japan.

It is spread by two kinds of snail – the *physopsis* and the *planorbis* – which feed on the undersides of water lily leaves. Tiny organisms develop in the intestine of both these snail types, which grow into long, microscopically thin bilharzia parasites that the snails finally eject. Having entered the skin, such organisms are carried through the blood stream to the bladder and rectum, causing damage to the kidneys and abdomen, and invading the lungs, heart and liver.

So there are many days when the riverside peasant feels anything but well, yet he must still tend his small plot of land to

[1] Also known as *schistosomiasis*. The actual parasites causing bilharzia are *schistosomes* and the male parasite lives *inside* the female.

produce vital food. This entails extensive cultivation, weeding, constant watering during the dry months, clearing away thick mud after the seasonal river floods, and always protecting the crop from animals, especially after daylight, when all-night vigils are necessary.

In Uganda, these Nilotic peasants are more fortunate because the Government pursues a vigorous health programme whereby doctors visit both large and remote village communities, but of course medical education involving prophylactic measures has never been achieved overnight and bilharzia and malaria are prevalent even around Paraa.

However, health teams are curing many cases of sleeping sickness, malaria and hook worm, while their continued efforts to eradicate tsetse flies and malaria carrying mosquitoes (the anopheles) must eventually bring improved conditions. By such methods, the *mwba* fly had already been exterminated in East Africa before Independence, and only bilharzia is the real stumbling block. At the moment there is no guaranteed cure and modern treatment still involves the use of drugs or poisons that often produce horrible side effects.

But visitors to East Africa are not running any risk of infection as long as they follow two golden rules: Take one daily tablet of the safe anti-malarial drug *Paludrine*, and keep your 'tootsies' out of deceptively inviting water. Two very simple rules that make a trip to the Nile or East Africa no more of a hazard to health than a picnic on Clapham Common.

We were now heading towards Murchison Falls themselves. Here the Nile was wide with a fast current and after making slow progress against the swirling water we eventually rounded a jungly headland to see the falls about 1,000 yards away. Then our motor stopped and immediately we were being carried downstream with an unpleasant spin. Using the stout boating pole as a 'stabilizer' did nothing to reduce our speed or keep us on any course.

C-R-A-S-H. We hit a mud bank island and shivered to a halt but we were still in one piece and apparently without structural damage. Crocodiles, disturbed by the collision, slid into the water.

Fortunately, our plight had been seen by another vessel. Although grateful for being rescued so promptly, we suffered the indignity of arriving back at the Paraa landing stage 'on tow' with a missing pole and a mangled propeller.

To really appreciate Murchison Falls they must be approached by road. The last few miles are a bumpy dust track cut through an area abounding with elephants. At the end of this track you leave the car and descend a rough path to huge flat rocks which border the River Nile as it surges from distant hills and boils along in a series of foaming, violent rapids.

It is possible to stand within six inches of these turbulent waters, but one false move or any unexpected wave backlash can bring disaster. Although the actual falls were now about 100 yards away, they were invisible from this point and only the trembling ground betrayed their presence.

From the rocks, a short, steep climb led to a high, narrow plateau jutting out over the impressive gorge where the mighty Nile is squeezed between a 19-foot wide cleft. Just think of the pressure created when you partly block a trickling tap or a hose-pipe with the finger and then imagine the same operation carried out on a scale millions and millions of times bigger. The result of such an operation would be spectacular to say the least, yet quite insignificant when compared with nature compressing the Nile through a mere 19-foot opening.

Colossal! Mighty! Stupendous! Awesome! If all the adverbs *of power* were strung together they could never describe Murchison Falls or capture the drama of watching the Nile rise as an angry, treacherous, whirling mountain of water that crashes into this narrow ravine and thunders from the other side with a blast that echoes for miles. Such pressure is far, far greater than any artificial power yet created by man.

Looking directly at this gigantic pressure hose, the clouds of spray seethed with repetitive patterns that exhibited a definite hypnotic fascination. I felt myself being drawn towards this eruption. I became obsessed with a feeling of buoyancy. A sensation of floating. Of drifting. Of complete weightlessness. My body was swaying. I was leaning forward as if being pulled by an invisible force.

Only a narrow strip of land separated me from those roaring

13.
A typical Kabale woman and her village, Uganda.

patterns of dense spray being propelled into the air with wild, infinite energy. Up and up and up; patterns which created spiralling columns of flashing marble that reached almost to heaven before they collapsed to form a magnesium bright firework cascade falling 140 feet to the calm Nile where hippos browsed and crocodiles slept.

Down below, the river was tranquil, but from up here the raging water was a turbine of unlimited power that generated fear, yet presented an irresistible attraction. And between this tranquillity and power lay only the clouds and the sky. The tranquillity was life. The power was death.

Such beauty can be ruthless and all-embracing. It is sometimes cruel, overwhelming, irresistible, demanding. Those who are unused to living close to nature are wise to savour her offerings delicately. Wise to show immense respect. Nature commands the greatest powers on earth and should not be taken lightly.

More than one visitor to Murchison has let this hypnotic exhibition of unrivalled grandeur drug their senses. Others, less perceptive, have taken foolish liberties. During the previous two weeks alone, Murchison had claimed the lives of three tourists who had inadvertently sacrificed themselves to beauty.

While the sun continues to shine, a rainbow spans the raging cataract of water as if to remind everyone that our Bible says God will never again flood the earth. And at the same time it is prudent to remember that 'in the midst of life we are in death'.

Go to Murchison. But go with reverence and humility. Go with caution and respect. Niagara, Victoria and a dozen equally celebrated falls may boom just as loudly, but I have found none of them to be as wild or unspoiled or so thrilling as Murchison. These falls in Uganda show nature at her magnificent best and they are among the most inspiring sights imaginable.

Humbly, I took my photographs and moved away from the edge of this precipice which separated me from tranquillity and power.

The tranquillity was life. The power was death.

14. A rare picture of the Vervet monkey sitting quietly, Tsavo, Kenya.

CHAPTER XXI

Strange Happenings

A State of Emergency – The new Republic – Important prefixes – Farmers – In Speke's diary – The Mountains of the Moon – Wine, Women and Song – A different 'call' – Mambas and cobras – False alarm – Magic stones – Poison antidotes – Stranger than fiction – Cow dung drugs – Children play with cobras – A secret which fades – Fish music – Less sceptical of African mysteries.

FROM Murchison Falls we drove southwest through the heart of Buganda, a former kingdom comprising a quarter of the country and which, only weeks prior to our arrival, was ruled by the Kabaka, a hereditary King. But the Kabaka was forcibly deposed and Buganda integrated with the rest of Uganda. Many lives had been lost in this political civil war, and a State of Emergency was still in operation but everything was almost back to normal.

War and politics are the abomination of mankind, yet frankly, if Uganda hoped to survive as an independent country, it was essential to integrate the kingdom of Buganda. I am not condoning the bloodshed, but unfortunately, while humanity remains uncivilized, most power squabbles cause violence. The other three less powerful kingdoms, Ankole, Toro and Bunyoro, had more or less accepted the need for unification, and with integration completed a special Constitution had just been formed to govern the new Republic of Uganda.

While travelling in Buganda, or even *reading* about the kingdom, to avoid absolute confusion it is essential to learn the meaning of a few important prefixes to the word *Ganda:*

*U*ganda is the name of the Republic.

*Bu*ganda is the name of a province which until recently was an independent, powerful kingdom.

*Mu*ganda is the name for a national of *Buganda*.

*Ba*ganda is the plural of *Muganda*.

*Ki*ganda refers to anything that has a specific connection with *Buganda* or the *Baganda*.

*Lu*ganda is the name of the language.

Ganda is a generic term for anything and everything relating to Uganda.

The Baganda number almost one and a half million and have a history that can be traced back for centuries; they are an extremely proud, reserved and cultured race of people. Much of their prosperity comes from growing cotton, coffee and maize. Practically every Muganda owns productive land, and many of Uganda's leading professional men are from Buganda.

When Speke discovered the source of the Nile he was particularly impressed by the Baganda, and again made an extra long note in his diary: 'The people were seen to be wearing neat barkcloth robes and over these went cloaks made from antelope skins which were sewn together as well as any English glovers could have pieced them; whilst their head-dresses, generally, were turbans, set off with highly polished boar tusks, stick charms, seeds, beads or shells, on their necks, arms and ankles they wore other charms of wood, or small horns stuffed with magic powder, and fastened on by strings covered with snake skin. I cut a poor figure in comparison with the display of the dressy Baganda,' he concluded.

The present day Muganda is still colourfully dressed, especially the womenfolk who wear vivid busuti dress in preference to western style clothing. Like the Masai, the Baganda also holds firmly to tradition, and in spite of development around Kampala, the people of Buganda would rather live in isolation.

Ruwenzori rises to the north of the Queen Elizabeth National Park. From a distance this great mountain range, thirty miles wide and stretching seventy miles between Lake Edward and Lake Albert, appeared to be an unusually dense formation of white clouds. But as we got nearer I could see inner patches of a distinct shape which were even whiter than the surrounding clouds. Disguised like pearls within an oyster, the patches soon became recognizable as the peaks and glaciers of Ruwenzori, the legendary Mountains of the Moon.

These wierd mountains on the equator are nearly always hidden by mist. The snowline starts at 14,500 feet and the highest peak, Margherita, is 16,763 feet. Ancient geographers like Herodotus believed the Nile to be fed by the ice melting from these glistening peaks.

Climbing to Margherita should be undertaken only by experienced mountaineers, and a guide is essential. However, Mount Speke at 16,042 feet does not offer any major difficulties. Bugoye, a village in the valley of the Mobuku River about thirty-five miles south of Fort Portal, is the usual starting point. It is a three-day climb to the Kitandara Hut at 13,000 feet, from where several peaks are easily attainable. At 10,800 feet, there is another hut at Nyameleju and this is within one day of the Ruwenzori glaciers. A fantastic circular tour is to Lake Bujuku, through the Scott Elliot Pass at 14,345 feet, over the Freshfields Pass, along several magnificent valleys and back via the Nyabitaba Hut; this takes seven days.

Between 9,000 and 6,900 feet, the slopes are covered with thick bamboo forest; below this level, for almost another 2,000 feet there are dense woods. At 5,000 feet, the aspect changes to rich, wide open grasslands and fertile, undulating landscape, through which we were now driving.

Here is some of Uganda's finest agricultural land, yet most of it was uncultivated, for this particular area is dominated by the Batovo. They are a tribe 600,000 strong known for their love of wine, women and song. Work is something which has never featured in the life of a Mutovo. (Mutovo is the singular of Batovo.)

Although many expert farmers from Buganda would gladly have settled around Ruwenzori, their hereditary King would allow nobody to leave his province. But with 'unification' completed, such restrictions no longer operate and it is probable that many Baganda will soon be working on this fertile land.

Wine, women and song. It was perfectly true; few Batovo

men were working in the fields. In the fields, yes; but working, no. Instead they lay beneath the cool shade of trees, drinking wine from flagons and occasionally playing a guitar. Most men were accompanied by a woman, often two, and sometimes three or even more! And what beautiful women. Indeed, Batovo girls are said to be among the most attractive and sensual women in the whole of Africa. I could well believe this, and certainly it is a fact that practically every singer, dancer, hostess and prostitute in the night clubs of Kampala is a Mutovo girl.

Where the fields *had* been cultivated, agricultural organizations were largely responsible, and tea was apparently the main commodity being grown. The tea seed was first introduced into Uganda during 1900 purely as a decoration for the Entebbe Botanical Gardens. Three years later more seed was obtained from India, and throughout different parts of the country spasmodic planting occurred at various intervals until 1924, when European planters suddenly took a great interest in producing tea and cultivated about 1,500 acres.

During 1929 Uganda exported 1,000 lb. of this product, and by 1970 the Uganda Tea Growers Association hopes to have 14,000 acres of tea plantations that will considerably increase their present export figure of 18,443,505 lb. This is an important development for Uganda, and African officials talk readily about the wonderful properties in 'a good old British cup of tea', forgetting that in the year 646 AD, Lo Tung, a Chinese poet of the Tang Dynasty said:

'The first cup moistens my lips and throat,
The second cup breaks my loneliness,
The third cup searches my barren entrail,
The fourth cup raises a slight perspiration,
The fifth cup purifies me,
The sixth cup calls me to the realms of the immortals,
The seventh cup – ah, but I can take no more.'

The 'call of the wilds' is rather different from a 'call of nature' and I had stopped in the fertile countryside around Ruwenzori and departed behind some bushes. Before my mission was completed I experienced the terror of having a slender, light green forest mamba,[1] about 7 feet long, slither past and almost wriggle over my shoes.

This was the first snake I had seen in Africa. While among the deadliest of species it is fortunately also the least aggressive, and rarely attacks without provocation, although any sudden movement on my part could have made the reptile strike by instinct. So, from necessity *and* fear, I froze into a motionless statue, and the mamba continued on its way.

Needless to say, my 'call' was completed with great speed. Only seconds after driving away, a black-necked 'spitting' cobra darted from the undergrowth and reared up to strike at the car. Had I encountered this vicious specimen instead of a mamba, I would have been in serious trouble because such a cobra attacks on sight.

Most East African snakes are venomous. They vary in size from huge pythons up to thirty feet or even more to the black worm snake which is no larger than a darning needle. Colouration generally consists of greens, browns and blacks, either immaculate or with distinctive patterns of stripes, mottles, spots and bars.

There are four main types of cobra. The black-necked 'spitting' cobra, the black-lipped cobra, the Egyptian cobra and Gold's forest cobra. The black-lipped species rarely exceed six feet but the other three frequently reach nine feet and occasionally more. The Egyptian cobra (a dark brown, tending towards grey or black, similar in colouration to the 'spitting' cobra) is common around the Murchison Falls. Gold's forest cobra (found only in the Queen Elizabeth National Park) has an enormous eye and a tail nearly a quarter of its total length.

It is unusual to see any of these reptiles by day, for all snakes see better at night and this is when their food is hunted. They are also deaf and perceive the approach of a body by vibrations conveyed through the ground. Because of this, it is wise to tread heavily; those who get bitten by a species like the forest mamba have usually made a silent approach, thereby giving the hidden snake too little warning of intrusion and frightening him into making a sudden attack.

In Nairobi there is a Snake Park, and while I was at the New

[1] Also known as a *green mamba*. The legendary *black mamba* does not exist.

Stanley the following report appeared in the *East African Standard:*

'Alarm and consternation hit the Nairobi Snake Park yesterday when a large car drove up to the park at speed and its driver rushed out to find the curator, Mr James Ashe.

'An unidentified snake had got into the car and was lying threateningly in the back, he told Mr Ashe breathlessly.

'The curator approached the car to investigate – watched by a large crowd which had gathered round the vehicle. Cautiously, he opened the door, found the snake, and caught it without difficulty.

'Identifying the reptile took only a few seconds. It was pronounced harmless but the driver had a long face when he drove off.

'Printed on the snake's body were the words *Made in Hong Kong*. It was made of rubber'.

Just about everybody is a target for African humour but their practical jokes are never really malicious. Somewhere in that 'large crowd' one person would have been splitting apart with laughter!

That night, while staying at an isolated rest house high in the Ruwenzori, we talked with an African biologist who, some years ago, had been working with conservationists in the Queen Elizabeth National Park. I related our experience with the forest mamba and cobra, but apparently, this area is one of the few regions in East Africa where snakes, for some reason unknown, are very prevalent and often seen during the day-time.

'Many of the land workers get bitten by the cobra, but now most of them have access to a magic mocobobo stone, few of them die,' the biologist told us.

Naturally, I wanted to know about the 'magic mocobobo stone', for here was an intelligent man, educated in England and with excellent professional qualifications, speaking of *magic stones*.

'No, no, I used the word *magic* just as Europeans acclaimed penicillin and tetracycline as the *miracle* or the *wonder* drugs. Let me rephrase that and say a *miracle stone*, although the natives still believe the mocobobo *is* magic. And why not? Most scientists have scoffed at the stone, and where any investigation has taken place the analyses show nothing unusual. Native superstition, they say. Believe you me, superstition or not, the stone works. I've used one myself on numberless occasions and every time a life has been saved.'

He pulled from his pocket a black stone about the size and shape of a small, flat Ever-Ready battery. It was dull black and looked solid, yet felt light and brittle with the texture of coke.

These stones are distributed to selected natives by the local missionaries and village chiefs. If anyone is bitten, a superficial incision is made where the snake's fang entered the skin; and once bleeding freely, a stone is placed on the wound, to which it clings like a leech. As even a hard knock will fail to dislodge the stone, patients can then easily follow their daily work routine.

After a few days the mocobobo drops off quite naturally, and having absorbed all the snake poison it is full of a bloody venomous liquid and no longer brittle or light. To reactivate its curative properties, the stone must be boiled in milk to release the absorbed poison and once cleansed the mocobobo floats on the surface of the milk where originally it sank.

The story was fantastic, but our biologist was adamant about the authenticity of this 'magic'. Yet again I recalled the words of Dr von Nagy spoken so many months ago: 'Things occur out here that are beyond normal comprehension and no matter how unbelievable the facts, then rest assured that at some time or another the thing has almost certainly really happened. Africa is a strange continent and there is nowhere else like it in the world.'

And after all, twenty years ago, doctors had laughed at the Masai who were said to cure terrible septic wounds by the liberal application of dung from their cows; but recently this dung has been found to contain various organisms similar to streptomycin and other modern antibiotics.

But there was no end to this biologist's repertoire of mysterious stories, and until 3.00 a.m. we were intrigued by his discourse. For example, in one special region, even the vicious cobras will never attack children, and toddlers often play with them quite safely. Once, our biologist could not avoid driving over a gigantic python which was surely all of 100 feet long and so thick and tough that his car was badly damaged. A friend of his was taming rhinos with recordings of symphonies, or stirring them into aggression by playing jazz. He knew of a baby girl reared in the jungle by the baboons.

'Tomorrow I will show you some of these things about which I talk and which you do not believe,' he concluded.

The biologist kept his word. After five hours' sleep, he was driving us through heavy rain forest in his Land Rover. Then, on foot, we made our way to a village community where two women both had a mocobobo adhering to their skin. One woman had been bitten in the leg, and the other on her left side just below the breast. Of course, it was impossible to *prove* that either of these women had actually suffered a deadly snake bite, but certainly the stone was clinging to them with an inexplicable firmness.

If these cases were genuine, could the explanation be faith? A 'mind' cure? An absolute conviction that the mocobobo would save them, just as their belief in the curse of a witch doctor could bring death! But this hardly explained the stones' leech-like characteristics. And where did the mocobobos come from? Could I take one away with me?

Apparently, faith played no role in this snake poison antidote, for the biologist told us he had used it successfully on people who knew nothing about the stone's reputation until they had fully recovered. Nobody had any idea from where the mocobobos originated, and certainly I would be forbidden to take a stone, for these were guarded no less stringently than the most valuable diamonds.

Even so, with all due regard for Dr von Nagy, I still remained very sceptical about the mocobobo stone. But later that day a great deal of this scepticism was dissipated when, in another village, we actually watched two very small children playing with a cobra as if this snake was no more harmful than a rubber toy. I wanted to take a photograph but the biologist restrained me.

'You are now seeing real magic. Watch silently. Do not intrude. Only the very young can charm these reptiles, and even their parents fail to understand how the act is accomplished. It is a childish game where secret powers are controlled unsuspectingly. A game they have played for generations in this same area and a secret which fades and dies from about the sixth year.

In the afternoon, we visited the man who was carrying out experiments to ascertain the effect of music on wild animals and his research was very similar to that which had been carried out by Hans Hass and Dimitri Rebikoff on the response of fish to music transmitted through water.

With aquatic life it has been established quite definitely that the transmission 'shock' waves produced by speakers relaying sound below the surface, have a distinct influence on fish behaviour, and waltz music – particularly the Blue Danube – attracts a great number of species, all of which swim around the transmission source in perfect rhythm.

Even ten years ago, several radio stations in America used to broadcast early morning experimental *Moo Moo Music* programmes for dairy farmers who were asked to collaborate with scientists to help prove that when certain kinds of music were played, cows gave their milk more freely and also produced greater quantities.

Now this research was being carried somewhat further by tackling wild animals, and around a waterhole there were four Goodmans Axiom loudspeakers which were encased in special housings and suspended from trees. Once again, the waltz had proved extremely popular with all the smaller animals, but rhinos and elephants preferred symphonies while buffaloes responded to the lighter classics. But whatever music was played – apart from jazz – the waterhole was always busier and the animals hardly ever squabbled. Yet, change from a symphony to something like *The Twelfth Street Rag* and in minutes the elephants would show great irritation and soon fights would develop between most of the animals, even those which were usually quite passive.

(It is interesting to note that Goodmans have led the field

in high fidelity speaker design for more than thirty-five years and are the largest manufacturers in Europe. They have now developed the *Subaqua 'P'* underwater speaker for permanent installation in test tanks and also the *Aqua-Chat,*[1] a new, versatile diver's communication system. Both these developments will help considerably in underwater research as well as having a wide range of uses for the amateur diver and marine biologist. My own books *Underwater Photography* and *Developments Underwater* (Swiss publication) both cover these subjects more fully.

For the human music enthusiast, Goodmans produce their *Audiosuite*, designed for every home and rated by experts to be among the very best stereo systems available today. This British organization therefore has the unique honour of bringing high fidelity reproduction not only to man, but also to fish in the tropical oceans as well as to the wild animals of East Africa!)

Similar experiments have been conducted at Treetops but these were never pursued scientifically and only operated spasmodically. From heresay, however, the results followed an almost identical pattern: symphonies for placidity and jazz for aggression. I wonder if any psychologist has thought of pursuing research into the 'pop' music of today, and its possible link with our youth problems where violence is becoming much more common?

After our strange experiences involving snakes, stones and symphonies at Ruwenzori we were far less sceptical and more fascinated than ever with the mysteries of Africa.

[1] A registered name of Partech (International) Limited.

CHAPTER XXII

Riches Beyond Value

Take note – Comparable with a lunar landscape – An elephant charge – Fear and concentration – Tropical Switzerland – Hard workers and happy – The sound of war – Gloom near the Congo – Mutilated bodies – A proud flag – Floods – Rough air – Twin valleys – Isolation – Three cheetahs – Ostrich farms – We dive and zoom – Our goodbye gesture.

ELEPHANTS Have Right of Way was the first sign we read in the Queen Elizabeth National Park. I had no intention of disputing the fact and we followed the track slowly, often having to stop while dozens of Jumbos wandered across our path.

This park covers 727 square miles of splendid scenery, a wide variety of rivers, swamps, lakes, open plains and deep forests. To the north lies Ruwenzori and to the southeast the Great Rift Valley wall. Over the vast grasslands grew giant cactus-type euphorbias; with their shape of an umbrella turned inside out they are unforgettable landmarks. On the drier plains, huge acacias and ivy-entangled fig trees provide cool shade for the wild life.

Unfortunately, our time was limited, but in spite of long, heavy rain periods we made a two-day circuit covering several hundred miles. We saw large herds of buffalo, impala, bush-buck, topi, Uganda kob[1] and a family of lions. We motored around a fascinating area where eighty-three extinct volcanic

[1] The strongest and most beautiful of the kob species, which is not found in Kenya or Tanzania. They have sleek red-gold coats and white markings on the legs and face. The females have no horns, but those of a fully grown male are more than one foot long and lyre-shaped. A buck generally weighs around 200 lb. and is remarkably agile, with amazing powers of scent and vision.

craters, all having precipitous sides rising from dense jungle, were crammed within only fifty square miles. From the air these craters are said to appear as a fantastic pattern of overlapping circles comparable with a lunar landscape.

It was while taking photographs here that we experienced our first elephant charge. Jill wanted a picture of me using a camera with an elephant in the far background. Although this particular elephant must have been 200 yards away he took a sudden and intense dislike to our activities. Out came his ears. Up went his trunk. He roared.

We gathered up the photographic equipment and jumped into the car. The engine fired immediately but Jumbo was now fairly close and gaining on us, ploughing through large bushes as if they were non-existent. At one stage he was immediately behind, thundering along at about 15 m.p.h.

I was driving as hard as possible on the twisting, narrow, rough track, dreading the thought of what would happen if we hit a large rock, got caught in the bog holes or had a puncture. To stall on a level crossing with an express train approaching must produce the same kind of fear. Crash – skid – swerve – brake – accelerate. I was sweating with the concentration and fright. Then Jumbo's speed began to fail. His sprint dropped to a fast walk, and the fast walk changed to a lumbering plod.

For a time we carried on in silence while regaining our composure. At length I broke the silence. 'Oh well, we've seen a jumbo jet before most people.'

This foolish little joke was a reaction to nervous stress, the same kind of forced joke people often make after having closely avoided a car accident; or perhaps like the small boy who whistles with a strained nonchalance before entering some dark cellar. So we both laughed as if *jumbo jet* was the funniest joke ever made!

Kabale is one of the most beautiful parts in Uganda; a tropical Switzerland. Everywhere there is a hill looming, with thatched roof houses either built far down on the slopes, or dangling precariously over the sides of a valley.

Wearing coloured headscarves, women helped their men

cultivate the fields even high up on the steepest mountain ranges, and it is said that you can always recognize a Kabale girl because of her powerful, muscular calves.

At one point the road zigzags through dense bamboo forest to 8,170 feet. Whenever we stopped to take photographs, or to admire a view, the hillsides echoed with the voices of women and children who chanted folk songs as they worked.

I doubt if any people in the world toil as hard as those around Kabale. From dawn until dusk they labour on every available square foot of arable earth, making terraced plantations which prevent soil erosion and create a striking network of orderly land contours that overlap and merge with intricate half-moon shaped patterns. It is hard, back-breaking perpetual labour but I doubt if any people in the world are happier, kinder or more gentle.

At Kisoro we reached the southernmost limits of our journey at the borders of Rwanda and the Congo. We stayed at the Traveller's Rest, once a flourishing guest house famous for its proximity to remote jungles inhabited by mountain gorillas. Now, with these gorillas finally claimed to be extinct and with constant war strife in the Congo (sometimes we heard the gunfire of the never-ending conflict), the Traveller's Rest was deserted. Its old furniture had gathered a thick layer of dust, shelves of empty bottles filled the unlit bar, and the whole atmosphere was pervaded by the gloom of disuse. Tanga had been a dying city, but Kisoro was already buried.

Yet a journey to the Traveller's Rest is worth while just to talk with the owner, a German who has lived there all his life in isolation with only one short visit to Kampala in the last fifteen years. Nobody knows more about the mountain gorilla or the Congo political situation than this German. He is an old man now, with memories authenticated by documents, diary notes and photographs that in themselves are already a part of history.

The Traveller's Rest is actually situated outside the border, on a stretch of 'no-man's land' that separates Uganda from the Congo and Rwanda. On our return, where the unsignposted road branched in three directions, we took great care over the decision. Some months previously, a party of students had accidentally driven into the Congo; they were later deposited at Kisoro with their livers ripped out, presumably because they were suspected of being mercenaries. It was a relief when we arrived at the familiar barrier where the red, yellow and black striped flag[1] of Uganda flew proudly in the wind:

'Hard have our people worked and prayed
To come unto our own proud right.
Behold the progress we have made –
Our homeland is a nation bright.

Behold the shining yellow sun.
Behold our freedom full and good.
Behold the blessings we have won.
Behold our land of brotherhood.

To keep our country strong and brave
We will work hard the years ahead.
Long may God let our Colours wave –
Our Black . . . Our Yellow . . . and Our Red!'

Uganda

It is approximately 500 miles, as the crow flies, from the Traveller's Rest to Kidepo National Park in the extreme north-east of Uganda. After many days of motoring through rapidly deteriorating weather conditions we eventually gave up the attempt by road. Rivers were flooding over parapets and many wooden bridges were in serious danger of collapsing. Each time we came to one of these rickety hazards I felt my legs become weak, although the Africans appeared to be far more confident. Having first tested the structure's rigidity *on foot*, they confidently proceeded to drive across with an overloaded 10-ton truck!

But we returned to Kampala, where a single-engined Piper Cherokee was kindly placed at our disposal by the Government of Uganda who were very keen for us to see their comparatively new conservation area of Kidepo.

Although the Cherokee's 180 h.p. engine gives a cruising

[1] In the centre of this flag is depicted a Crested Crane, which is the 'symbol' of Independent Uganda just as the Scarlet Ibis bird has become the emblem of Trinidad.

maximum of 152 m.p.h., exceptionally turbulent conditions forced us to fly at little more than 110 m.p.h. with frequent course alterations to avoid widespread thunderheads. It was a rough trip through intermittent thick cloud and heavy rain, but the Cherokee handles particularly well in such conditions and at Kidepo the sun welcomed our landing on the grass strip.

This national park lies in a basin almost encircled by mountain ranges. On the west lies the Napore range (Lonyili at 7,500 feet), on the east the Zulia range and Turkana escarpment (Zulia reaches 7,047 feet) and to the south is the Morungole range with Morungole being 9,022 feet, Kawalakol 7,421 feet and Lomej 6,516 feet. For thirty miles, the Park's northern border is separated from the Sudan by an international boundary line, although large bands of armed Sudanese cattle raiders still make frequent swoops upon the area to herd off livestock belonging to the local Karamoja people.

The 500-square mile of Kidepo National Park was created in March 1962.

In the twin valleys of Narus and Kidepo 3,000 and 4,000 feet respectively, the vegetation is mainly acacia woodland and savannah, while on the higher slopes there are montane forests. From October until May this is an arid region, with the river from which the Park takes its name being only a dry sand bed. Even after torrential rains[1] the water is completely absorbed by the black soil and although the Kidepo River may be a raging torrent for several hours, it quickly reverts to sand again.

Kidepo is the most isolated of any national park in East Africa, and it will inevitably remain 'off the map' for the majority of tourists until there are better roads, and/or a regular light aeroplane service, despite the excellent accommodation facilities.

The facilities include ten double bandas each complete with private bathroom having hot and cold showers. Food is self-prepared in a central banda but a cook and houseboy is also available if required; there is a shop for non-perishable goods. A generator provides constant power and the kitchen is

[1] Most of Uganda's rain falls between April and September; the yearly average is 35 inches for the wettest area and 15 inches in the Kidepo Valley itself.

15. Seronera Camp, Serengeti, Tanzania.

A male and female rhino, Tanzania.

equipped with refrigerators and a deep freeze. Firewood and water is available for the camper. Land Rovers and guides can be hired and there is a radio telephone link with Kampala.

Having given notice of our arrival by making a dummy run over the grass airstrip, the Park Warden and his wife (Mr and Mrs Ross) were waiting to meet us when we landed; a young, charming, recently married couple, both born in East Africa and both with an equal passion for conservation.

The Valley offers them no parties, no shops, no radio, no artificial entertainment of any kind, no ladies' hairdresser and not even the simplest of everyday pleasures. Yet Kidepo gives their lives a worthy purpose. It gives them a vital spark, increased perception and an affinity with the unknown powers of creation. And most of all it gives satisfaction by the partial fulfilment of a spiritual need that I am certain lies in every person. These are intangible reflections which remain unobtainable through wealth and power. Such commodities are not for sale and they are beyond value.

A million pounds would surely fail to attract Mr and Mrs Ross from their mountains, their animals and their remote valley. They are wise enough to know that Kidepo offers greater riches than any bank balance.

During our stay the rains kept away from Kidepo and we were able to explore by Land Rover, at one time driving along the Sudanese border for several miles.

On the last evening we saw our very first cheetahs; three of them, on the run, looking like extra long-legged greyhounds covered in leopard skins!

These animals combine the best characteristics of both the dog and the cat, although much of the cat's ability to climb has been lost. Cheetahs lack the natural savageness of a leopard and will rarely attack an aggressor even if wounded. They are easily tamed and make affectionate pets, having no vicious hooked claws[1] to cause accidental injury.

Disturbed by the Land Rover, the cheetahs sprinted away,

[1] The claws of a cheetah cannot be retracted and are blunt like those of a dog.

16. An elephant flaps its ears with annoyance, Tsavo, Kenya.

probably topping 30 m.p.h., but not reaching the 70 m.p.h. which they can achieve for 250 yards while hunting. After stalking a herd of antelope to within 100 yards she relies on her tremendous speed to launch an attack, covering the 100 yards in about three seconds and overtaking gazelles even when they are fleeing at 40 m.p.h. A cheetah never kills unless hungry.

As cheetahs are much rarer than the lion (there are possibly no more than 750 left in East Africa), this brief sighting was very lucky, and our last evening at Kidepo was therefore spent in jubilation and sadness.

The morning came all too soon and we both rose soon after dawn. Outside our banda the plains were dotted with antelope, gazelle, bushbuck, dik-dik and scores of ostrich.

Ostriches are the largest of living birds and to watch the female run is suggestive of an elderly woman who was once a cancan girl and, although grown somewhat ungainly, she still maintains a little grace and delights in tossing her feathers to give an undignified backside view!

An ostrich egg weighs around 3 lb.[1] (equivalent to twenty-four hens' eggs). The Rosses recently found a deserted nest containing several eggs from which batch they managed to incubate two of them. Unfortunately, a lion took one of these babies, but *Ziggy* survived and was a ten-month-old ostrich with whom we became great friends, and who we think has made a charming study for the cover of this book.

Between 1903 and 1930 almost 1,000,000 ostriches were killed for their plumage, but today these feathers are not so fashionable and the demand is easily met by several ostrich farms where 50,000 birds are kept. From the age of six months the soft feathers can be taken regularly until about the sixteenth year; then the ostrich is killed, its skin sold to the leather industry, and its bones go for fertilizer and cattle food.

The income from each bird totals around £1,300. I hope Ziggy realizes his good fortune.

In the wilds an ostrich lives for forty-five years or more, and being able to out run most predators this earth-bound bird[2] has no real enemies apart from man.

[1] Eggs from the extinct *Madagascar ostrich* weighed 18 lb.
[2] Birds weighing over 40 lb. are unable to fly.

Those who wish to experience a National Park that has remained truly wild, remote and unspoiled, should make an effort to reach the Kidepo Valley. When a letter is posted at Kidepo, the local postman takes it to the nearest collection point by *cycling* through the bush for 125 miles! Here is an enormous area of East Africa where even the most hardened city man could not remain impassive to nature and the stimulus of her great outdoors.

Already Kidepo was 1,000 feet below us. We swept round in a goodbye circuit, put our nose down, gently pulled out of the dive and with an air speed of almost 200 knots roared over the airstrip at less than fifty feet.

Ross and his wife were waving. Then, on full power, we zoomed steeply to over 3,000 feet like an express lift. Momentarily the stall warning light flickered as we reached the climax of our ascent before levelling off to set a course for Kampala.

This is how *not* to fly, but as we were miles from any town or air traffic, perhaps the showy finale may be excusable. It was no more than a flourish beneath the signature on some important document, a way of expressing emotion with the tools that lay immediately to hand. It was our personal farewell to Kidepo and the wild glory of Africa which soon we should be leaving.

CHAPTER XXIII

No Endeavour is in Vain

'The wanderer is an eternal being who must forever circle the globe in search of food for meditation and the overwhelming interest in the company of his fellow men.'

Muhy-d-din ibn' Arabi.

THE British United VC 10 lifted from the runway at Entebbe.

Over Lake Victoria. Over Kampala which is built on seven hills like Rome and Lisbon. Over the wide avenues and spacious parks. Over the bubbling Nile where hippos and crocodiles reign supreme. Over the arid plains that vultures are always circling. Over green valleys, the home of elephants and buffaloes. Over dark jungles. Over acacia forests where antelopes graze and lions roam. Over snow-capped mountain peaks.

Over the clouds. Higher and higher. Then Africa disappeared and only a bed of cotton wool lay below us. Africa had gone. I felt sad. This was the real goodbye and there would be no farewell zoom in our VC 10.

We shall never forget Africa, its ever-changing landscape, warm-hearted people, azure sky and fascinating wild life. Each day she had offered us a new, wonderful discovery, discoveries that blazed a trail of enchantment. Although the initial thrill of such discoveries can never be recaptured, we hope *Wild Africa's Silent Call* may persuade others to follow the same trail, and those who do will have their reward.

Africa is not a paradise for weak-willed escapists, not a haven for the disillusioned, not a Mecca for the professional failures, not an easy retreat for those who wish to hide from life.

But Africa offers its very soul to those who will expend a little energy, enthusiasm and tenacity in making their discoveries. As Sir Thomas Browne once wrote: 'We carry within us the wonders of Africa we seek without us.'

What had our own journey achieved? We did nothing of special merit except to persevere with enthusiasm and to create a document that we hope will inspire many readers to think seriously about the world we live in – about ourselves, about wild life, and whether humanity stands for *conservation* or *extermination*.

I think our biggest gain was the achievement of personal satisfaction for a few short weeks. As the great African pioneer, Stanley, once quoted from Longfellow:

> 'No endeavour is in vain;
> Its reward is in the doing,
> And the rapture of pursuing
> Is the prize the vanquished gain.'

APPENDIX I

A Quick Reference to East African National Parks

UGANDA

1. *Murchison Falls* (see pages 181 to 193).

This National Park is named after the awe-inspiring Murchison Falls, where the swirling waters of the Victoria Nile thunder through a 19-feet wide gap before plunging 140 feet to the river below.

This park stretches from the banks of the river on either side. Huge herds of elephant, buffalo and many of Uganda's species of antelope can be seen. The most popular game viewing trip is from a launch which takes visitors up the Nile for several miles. The river is teeming with crocodile and hippo.

The bird life includes such species as the goliath heron and the whale-headed stork. Nile perch and tiger fish can give anglers great excitement.

2. *Kidepo Valley* (see pages 207 to 211).

The newest National Park in Uganda, Kidepo stands to the extreme northwest of the Karamoja District. It covers 500 square miles in one of the most scenic parts of Uganda. There is a landing strip and a self-service lodge in the Park which has an abundance of game. It is served by roads from Uganda and Kenya via Kitale.

3. *Queen Elizabeth* (see pages 204 to 206).

Bordering the romantic Mountains of the Moon to the north and stretching to Lake Edward and Lake George, this is Uganda's biggest wild life sanctuary.

Scenery varies from open plains and forest to a fascinating area containing more than eighty extinct volcanic craters in only fifty square miles.

The waters of the Kazinga Channel joining Lake Edward to Lake George hold a huge hippo population and a tremendous variety of birds.

Lions can often be found and herds of elephant, buffalo and antelope are common. Chimpanzees live in the forested edges of the Park near the Congo border on the Ishasha River.

TANZANIA

4. *Serengeti* (see pages 136 to 141).

Tanzania's foremost national park, this attracts thousands of tourists to see the great seasonal migrations of game, particularly of wildebeest and zebra which move from the central grasslands to the permanent water holes in the western sectors of the Park. Although these migrations are perhaps most spectacular after the long rains in May and June, these huge herds of animals do, in fact, rotate through fresh grazing grounds during the whole year.

The plains are also famous for huge prides of lion. Leopard and cheetah are also fairly common. There are now a considerable number of elephants living in the Park, many of which have moved in permanently to take refuge from encroaching cultivation to the west of Serengeti.

5. *Lake Mayanara* (see pages 112 to 117).

Standing at the foot of the Great Rift Valley wall, and only about seventy miles from Arusha, the scenery here is varied and beautiful. The park is famous for its collection of lions which habitually rest in the thorn trees during the day.

6. *Arusha* (see pages 28 to 50).

This is a new Park taking in the Momella Lakes, Ngurdoto Crater and Mount Meru. Game is plentiful and there are some magnificent views of Mount Kilimanjaro and Mount Meru.

7. *Ngorongoro* (see pages 118 to 135).

This famous extinct crater, one of the world's largest, offers unusual wild life and scenic spectacles. Huge herds of antelope graze on the floor of the ten-mile wide crater. There are more than fifty lions. Elephant, buffalo, hippo, leopard and cheetah can also be seen.

Only a few miles away is the archaeological site of Oldupai Gorge, where several *Homo habilis* and *Zinjanthropus* skulls have been found.

8. *Ruaha*

This National Park stands to the west of the Great Ruaha River and spreads over more than 5,000 square miles. All the varieties of big game animals can be seen here.

9. *Selous*

This is said to be East Africa's largest game reserve although at the moment it is undeveloped. Situated astride a complex of numerous rivers, this may become Tanzania's leading attraction.

10. *Mikumi*

Lions, elephant, buffalo, hippo, giraffe are all found in this 450-square mile Park to the north of Selous and within reach of Dar es Salaam.

KENYA

11. *Mara* (see pages 144 to 145).

Situated on Kenya's southern border with the famous Serengeti, the Masai Mara Reserve has become an increasingly popular attraction for safari circuits. Huge herds of buffalo roam these rolling grasslands and there are several prides of lion.

12. *Amboseli* (see page 26).

The most popular haunt of amateur and professional photographers. Here, a variety of game which includes large herds of rhino and elephant, can be seen against the backdrop of Tanzania's Kilimanjaro.

13. *Tsavo* (see pages 56 to 66).

8,000 square miles of rugged Africa which is one of the greatest remaining wild life strongholds in the world. Recently the elephant population has increased to over 20,000. There are also large numbers of rhino and buffalo. Other attractions include Mzima Springs, Roaring Rocks and the Aruba Dam.

14. *Nairobi National Park*

A compact area of forty-four square miles which is just a ten-minute drive from Nairobi. There are large prides of lion and big herds of zebra, wildebeest, gazelle, impala, waterbuck and ostrich.

15. *Lake Nakuru*

Situated in the Great Rift Valley among the hills two miles south of Nakuru this is widely considered to be the finest bird sanctuary in the world. It is famous for the half million flamingoes which live along its shores and additionally more than 300 species of bird have been classified on the lake.

16. *Mount Kenya* (see pages 170 to 172).

Almost the entire area of this majestic mountain, from the dense forests at its base to the snow covered twin peaks at over 17,000 feet, are included in the Mount Kenya National Park. This is Kenya's highest mountain and sitting astride the equator it provides spectacular scenery. Land Rover and muleback rides can be arranged through the forest and on to the moorland slopes below the Lewis Glacier.

17. *Aberdare* (see pages 155 to 169).

Treetops Hotel is the most famous feature of this mountain park. The 228 square miles abound with elephant, buffalo, rhino and the rare bongo as well as many other forest animals. There are waterfalls, hundreds of feet high, and many fast-running streams. Permission for trout fishing can be obtained.

18. *Marsabit and Samburu*

Over 100 miles apart and both situated in the remote and arid expanses of northern Kenya. Special permission must be obtained from the authorities to visit Marsabit.

19. *Meru*

This has only recently acquired national park status. Its creation earlier as a reserve was due largely to the inspiration of George and Joy Adamson, foster-parents of the famous lioness Elsa. Many rivers in the area are well stocked and anglers can mix their game viewing with excellent fishing. This Park should not be confused with Meru in Tanzania.

APPENDIX 2

Notes on the Major Animals Mentioned in this Book

ANT-BEAR. *Orycteropus afer lademanni.*

Swahili name: *Muhanga.*

Length: 4 to 5 feet.

Special characteristics: The long, narrow tongue which is inserted into ant-holes measures up to 20 inches. The tail is 17 to 24 inches long.

Weight: Up to 140 lb.

Additional notes: Very rarely seen during the day.

BUFFALO. *Syncerus caffer caffer.*

Swahili name: *Mbogo* or *Nyati.*

Height: 5 feet at the shoulder.

Horns: 28 to 42 inches.

Weight: Up to 2,000 lb.

BUSHBUCK. *Tragelaphus scriptus delamerei.*

Swahili name: *Pongo.*

Height: 33 inches at the shoulder.

Horns: 12 to 16 inches. Up to 22 inches. Usually only developed in the male but occasionally also in the female.

Weight: 90 to 170 lb.

CHEETAH. *Acinonyx jubatus raineyi.*

Swahili name: *Duma.*

Height: 2½ to 3½ feet at the shoulder.

Length: 6 to 7 feet. Up to 7 feet 9 inches.

Special characteristics: The tail is almost 3 feet long.

Weight: 90 to 140 lb.

DIK-DIK. *Rhynchotragus guentheri smithii.*

Swahili name: *Dikidiki.*

Height: 14 to 18 inches at the shoulder.

Horns: Very small and only present in males. Hardly visible.
Weight: 8 to 9 lb.
Additional Notes: Usually seen in couples or singly.

ELEPHANT. *Loxodonta africana.*
Swahili name: *Ndhovu* or *Tembo.*
Height: 10 foot 9 inches at the shoulder. Up to 13 feet.
Tusks: Up to 11 feet 5½ inches along the curve. Weight 50 to 130 lb. but up to 235 lb.
Weight: 4 to 6 tons.

GIANT FOREST HOG. *Hylochoerus meinertzhageni meinertzhageni.*
Swahili name: *Nguruwe.*
Height: Up to 32 inches at the shoulder.
Tusks: Up to 12 inches long.
Weight: Up to 299 lb.

GIRAFFE. *Giraffa camelopardalis tippelskirchi.*
Swahili name: *Twiga.*
Height: 8 to 12 feet at the shoulder. Up to 15 feet at the horns in females and up to 18 feet for males.
Weight: Approximately 1 ton.

HIPPOPOTAMUS. *Hippopotamus amphibius kiboko.*
Swahili name: *Kiboko.*
Height: 4 feet 5 inches to 5 feet at the shoulder.
Length: Up to 14 feet.
Tusks: From 20 inches to 30 inches but up to 41 inches.
Weight: 4 to 5½ tons.

HYENA (SPOTTED). *Crocuta crocuta germinans.*
Swahili name: *Fisi.*
Height: 2 feet 2 inches to 3 feet at the shoulder.
Length: 5 feet to 6½ feet.
Weight: Up to 180 lb.

IMPALA. *Aepyceros melampus suara.*
Swahili name: *Swala Pala.*
Height: 32 to 40 inches at the shoulder.

Horns: 29 to 30 inches and up to 33 inches along the curve. In males only.
Weight: 60 to 75 lb.

JACKAL. *Canis mesomelas elgonae.*
Swahili name: *Bweha.*
Height: 2 feet to 2½ feet at the shoulder.
Length: 3 to 3½ feet.
Special characteristics: The tail is up to 15 inches long.
Weight: Up to 22 lb.

LEOPARD. *Panthera pardus pardus.*
Swahili name: *Chui.*
Height: Up to 30 inches at the shoulder.
Length: 7 feet 6 inches to 8 feet and over.
Weight: 100 to 130 lb.

LION. *Panthera leo massaica.*
Swahili name: *Simba.*
Height: 3 feet 4 inches to 4 feet at the shoulder.
Length: 9 feet to 9 feet 4 inches but up to 10 feet.
Weight: 400 to 500 lb.

RHINOCEROS. *Diceros bicornis bicornis.*
Swahili name: *Faru.*
Height: 5 feet 6 inches to 5 feet 10 inches at the shoulder.
Length: 11 feet.
Horns: 18 to 20 inches but up to 43 inches.
Weight: 2 tons.

THOMSON'S GAZELLE. *Gazella thomsonii thomsonii.*
Swahili name: *Swala Tomi.*
Height: Up to 2 feet at the shoulder.
Horns: 12 to 15 inches but up to 17 inches.
Weight: 50 to 65 lb.

TOPI. *Damaliscus korrigum.*
Swahili name: *Nyamera.*
Height: 45 to 50 inches at the shoulder.
Horns: 12 to 15 inches but up to 19¼ inches. Present in males and females.
Weight: 260 to 300 lb.

WARTHOG. *Phacochoerus aethiopicus aeliani.*
Swahili name: *Ngiri.*
Height: 26 to 30 inches at the shoulder.
Length: 2 feet 10 inches to 3 feet 2 inches.
Tusks: Up to 25 inches.
Weight: 175 to 220 lb.

WATERBUCK. *Kobus ellipsiprymnus kuro.*
Swahili name: *Kuro.*
Height: 45 to 56 inches at the shoulder.
Horns: 24 to 26 inches but up to 33 inches.
Weight: 385 to 480 lb.

WILD DOG: *Lycaon pictus lupinus.*
Swahili name: *Mbwa Mwitu.*
Height: 2 feet to 2 feet 8 inches at the shoulder.
Length: 4 feet 10 inches.
Special characteristics: The tail is 14 inches or more long.
Weight: 50 to 85 lb.

WILDEBEEST (GNU). *Connochaetes taurinus albojubatus.*
Swahili name: *Nyumbu.*
Height: Up to 52 inches at the shoulder.
Horns: 25 to 28 inches but up to 30 inches and more. Present in both sexes.
Weight: 500 to 600 lb.

ZEBRA. *Equus burchelli granti.*
Swahili name: *Punda Milia.*
Height: Up to 4½ feet at the shoulder.
Length: 8 feet.
Weight: 450 to 750 lb.

NB. All these statistics represent an *average* figure.

APPENDIX 3

Useful Addresses and Information

FOR SAFARIS IN TANZANIA

Tanzania National Tourist Board,
111, Grand Buildings, Trafalgar Square,
London, W.C.2. Telephone: 01-930 5701

Tanzania National Tourist Board,
Head Office, P.O. Box 2485,
Dar es Salaam, Tanzania. Telephone: 24091

Tanzania National Tourist Board,
Caxton House, Kenyatta Avenue,
Nairobi, Kenya. Telephone: 26888

Tanzania National Parks,
P.O. Box 3134,
Arusha, Tanzania. Telephone: 2335

United Touring Company,
P.O. Box 3173,
Arusha, Tanzania.

FOR SAFARIS IN KENYA

Kenya Tourist Office,
318, Grand Buildings, Trafalgar Square,
London, W.C.2. Telephone: 01-839 4477

Kenya Tourist Office,
Victoria Street,
Nairobi, Kenya.

United Touring Company,
P.O. Box 2196,
Nairobi, Kenya.

Block Hotels Limited,
P.O. Box 75,
Nairobi, Kenya.

FOR SAFARIS IN UGANDA

The Embassy of Uganda,
58, Trafalgar Square,
London, W.C.2.

The Ministry of Information Broadcasting & Tourism,
Kampala, Uganda.

United Touring Company,
P.O. Box 167,
Kampala, Uganda.

FOR GETTING TO EAST AFRICA

British United Airways Limited,
Portland House,
Stag Place, London, S.W.1.
Reservations Telephone: TATe 9066

United Touring Company,
Berkeley Square House,
Berkeley Square,
London, W.1.

For those people who may be flying from the Continent I thoroughly recommend Lufthansa, Germany's National Airline which covers most of Western Europe and also operates a regular service to Kenya, Tanzania and Uganda. The Germans are fine technicians, excellent mechanics and make natural pilots. Throughout the world, Lufthansa's striking motif, a blue condor bird against a yellow background, is a renowned symbol for efficiency and safety.

Swissair, another leading European airline is also highly recommended, although at the present time they only operate a service to Nairobi.

Should readers have any queries the author will be very pleased to answer them promptly if they write to him c/o Allen and Unwin Limited enclosing a stamped addressed envelope. Any unfortunate delay in replying will simply mean that Derek Townsend is 'on safari' and cannot be contacted.

Index